# BACK FIELDS

PRACTICAL PREP FOR EXAMS IN E&M

DANIEL A. MARTENS
YAVERBAUM

Tattenwitch Editions, being an imprint of
Law Four Press
Brooklyn, NY



ISBN-s: 978-0-9985847-9-9

In a letter to Lord Richard Bentley,
Trinity College, University of Cambridge,
25 Feb. 1693:

"... That gravity should be innate, inherent and essential to matter... is to me so great an absurdity that I believe no man who has in philosophical matters a competent faculty of thinking, can ever fall into it."

— NEWTON

# CONTENTS

# PART I

SUMMER 2015

1

# FULL-TEXT OF FINAL EXAM SU15: BLANK

# FINAL EXAM:
# PHYSICS 204, JULY 17, 2015.

~,Charge, Flow, Field, Flux & Lu},.

JOHN JAY COLLEGE OF CRIMINAL JUSTICE,
THE CITY UNIVERSITY OF NEW YORK

DANIEL A. MARTENS YAVERBAUM

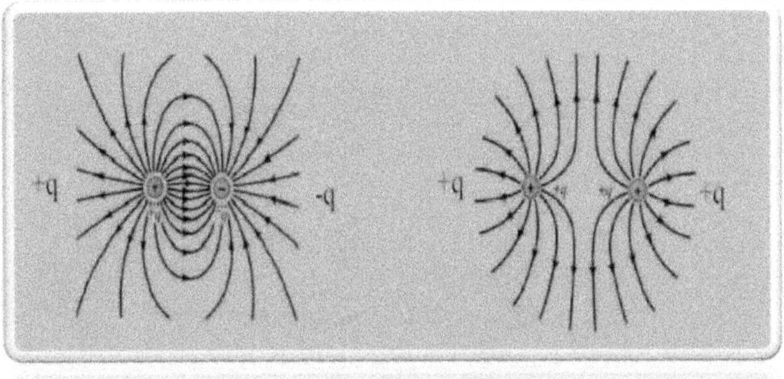


**Name:** ______________________________

**Section #:** __________

SCORE: __________

## SOME USEFUL RELATIONS:

1) $\oint \vec{E} \cdot \overrightarrow{da} = \frac{q_{(enc)}}{\varepsilon_0}$.

2) $\oint \vec{B} \cdot \overrightarrow{da} = 0$.

3) $\oint \vec{E} \cdot \overrightarrow{dl} = -\frac{d}{dt}\int \vec{B} \cdot \overrightarrow{da}$.

4) $\oint \vec{B} \cdot \overrightarrow{dl} = \mu_0 I_{(enc)} + \mu_0 \varepsilon_0 \frac{d}{dt}\int \vec{E} \cdot \overrightarrow{da}$.

5) $\vec{E} = \frac{1}{4\pi\varepsilon_0}\frac{q}{r^2}\hat{r}$.

6) $\vec{F}_E \equiv q\vec{E}$.

7) $\vec{B} = \frac{\mu_0}{4\pi}\frac{q}{r^2}\vec{v} \times \hat{r}$.

8) $\vec{F}_B \equiv q\vec{v} \times \vec{B} = \vec{I}l \times \vec{B}$.

9) $V_a - V_b \equiv \int_a^b \vec{E} \cdot \overrightarrow{dr}$.

10) $C \equiv \frac{Q}{\Delta V}$. (Note: This "C" is capital.)

11) $I \equiv \frac{dq}{dt}$

12) $I = \frac{\Delta V}{R}$.

13) $\mathcal{E} - IR - \frac{Q}{C} = 0$. (Capital "C".)

14) $c \approx 3\,x\,10^8\ m/s$.

15) $n \equiv \frac{c}{v}$. (Lower-case "c".)

16) $n_1 \sin\theta_1 = n_2 \sin\theta_2$.

17) $\frac{Yd}{L} = \frac{n\lambda}{2}$

SCORE: ____________

18) $sin^2\theta + cos^2\theta = 1.$

19) $\bar{v} \equiv \frac{\Delta x}{\Delta t}.$

20) $\sum \vec{F} = m\vec{a}.$

21) $F = -Kx.$

22) $x = Acos(\omega t + \phi).$

23) $\omega = 2\pi f.$

24) $f = \frac{1}{T}.$

25) $KE = \frac{1}{2}mv^2$

26) $PE_{elastic} = \frac{1}{2}Kx^2$

27) $v = \lambda f.$

28) $v = \sqrt{\frac{T}{\mu}}.$

29) $\frac{\partial^2 y}{\partial t^2} = (v^2)\frac{\partial^2 y}{\partial x^2}.$

30) $\epsilon_0 \approx 8.85 \, x \, 10^{-12} \, \frac{C^2}{Nm^2}.$

31) $\mu_0 \approx 1.26 \times 10^{-7} \, \frac{N}{A^2}.$

32) $m_e \approx 9.11 \, x \, 10^{-31}$ kg.

33) $m_{(NEUTRON)} \approx 1.67 \, x \, 10^{-27}$ kg

34) $q_e \approx 1.60 \, x \, 10^{-19} \, C.$

SCORE: ___________

## I. E-FIELDS FROM ***POINT CHARGES*** (20 PTS).

Two ***point-charges*** of differing magnitudes and are held stationary in an enormously spacious x-y plane.

A researcher places an instrument called a 'field detector' at the point (-5,+12). She is interested in measuring the electric field at that precise location.

The two point charges are as follows:

| Name | Charge | x-Coordinate | y-Coordinate | Ordered Pair |
|---|---|---|---|---|
| $Q_1$ | $+5 \times 10^{-10}$ Coulombs | +5 meters | 0 meters | (5,0) |
| $Q_2$ | $-12 \times 10^{-10}$ Coulombs | 0 meters | -12 meters | (0,-12) |

**Location of Interest:** (0,0)

Note: All coordinates are measured and given in **meters** (not centimeters); similarly, the (enormous) charge magnitudes are in whole **Coulombs** (not micro-Coulombs).

Also Note: If you wish, you are permitted and encouraged to approximate the electrostatic constant as:

$$K_e \approx 10 \times 10^9 \approx 1 \times 10^{10} \frac{Nm^2}{C^2}.$$

* * *

a) Draw a neat and clear sketch of the situation, as you understand it. Your sketch must express a clear decision as to which directions are designated by + and – on each axis (3 pts).

b) For *each* of the two individual point charges, draw an approximate field line diagram – each drawn as it would look IF the other charge did NOT exist. For BOTH, however, obey the following convention: For every $1 \times 10^{-10}$ Coulombs of strength that generates a field, at least one more field line should appear in the field line diagram (3 pts).

SCORE: __________

c) Compute the Electric Field as measured at the Point of Interest (-5,+12).

That is:

i. In Newtons/Coulomb, determine the ***electrostatic field*** **magnitude** at this location of interest (5 pts).

ii. Using degrees **and** points of the compass (such as "20 degrees South-West"), express the **direction** of the ***electrostatic field*** at this Point of Interest (-5,+12) (3 pts).

d) Staying approximately consistent with whatever scale, style (etc.) was involved in your diagrams for (b), above, try now to bring the two graphics together into one visual ***superposition*** of the situation: That is, draw one field line diagram for the net influence exerted by this pair of point charges. In your mind, start bringing the two pictures increasingly close together... Since lines cannot, however, cross nor disappear, allow denser regions of lines from one charge to gently and smoothly 'bend back' (or forward) the lines from another – doing as much as you can to preserve simplicity and symmetry where applicable (3 pts).

e) Assume that a somehow isolated and highly condensed Helium Nucleus (consisting two protons, two neutrons and no other measurably significant entities) is introduced and held gently at the point (-5,12). Assume, further, that any uncharged particles trapped in that nucleus simply 'go along for the ride' whenever the charged particles are motivated to travel. The nucleus is then released.

Compute the nucleus 's initial instantaneous acceleration. Provide precise **magnitude** AND **direction** (3 pts).

SCORE: ___________

## II. Gauss's Law (25 pts).

A huge but imaginary sphere is drawn with its center at the origin of the coordinate system used in Problem I, above.
The radius of the sphere is $r$ = ***15 meters***.

**ALL INFORMATION given or found in the above Problem I STILL APPLIES.**

The huge and imaginary sphere contains both of the charges given in Problem I.
And nothing else.

a) *Draw this situation* as you understand it (2 pts).

(This is not meant to be a trick; the drawing will just help clarify.)

b) In $\frac{Newtons \cdot Meters^2}{Coulombs}$, find the total amount of ***electric flux*** that passes through the surface of this imaginary sphere (2 pts).

Your answer should be a number, expressed in the units mentioned directly above.

c) The two point charges from Problem I are replaced with an extremely long and straight ***wire*** (i.e.: LINE) that is net-positively charged; this straight line wire stretches through the points (0,+5) and (-12,0) and beyond.

The charge density in the wire is constant and expressed as follows, where $L$ refers to length measured in meters and $Q$ is charge measured in Coulombs:

$$\lambda \equiv \frac{+Q}{L}$$

*Your first large goal, after a series of smaller 'build-up' questions to follow on the next page, will be to compute the electric field as a function of perpendicular distance,* ***r****, from this line of charge.*

i. Draw the field line diagram produced by this long line of positive charge (2 pts).

ii. In a sentence or two of English, explain why the sphere drawn above will not be the most convenient shape for computing ***electric field*** at $r$ (3 pts).

SCORE: ___________

iii. Choose and draw a Gaussian (closed) surface surrounding some portion of the line—a surface that will best help you compute the field at $\boldsymbol{r}$ (2 pts).

iv. Starting with Gauss's Law and proceeding through as many clear and verbally explained steps as possible,

find $\boldsymbol{E}$ as a function of (perpendicular distance) $\boldsymbol{r}$
(given $\boldsymbol{\lambda}, \boldsymbol{\varepsilon_0}$)
from this line of charge (5 pts).

This answer will not be a number.

v. Now assume that $\lambda \equiv 5\frac{Coulombs}{meter}$.

In VOLTS, Find the ***Electric Potential Difference*** ('Voltage'!) that exists
(as long as that line remains net-charged)
between the point **(5,-6)** and the point **(10,-18)**.

This is not a joke. Nor a trick.
Hint: Recall and deploy the definition of ***potential difference*** (5 pts).

vi. Batteries and power supplies are generally used to supply ***Electric Potential Difference*** ('Voltage') to circuits, thereby allowing us to operate electric and electronic devices. If, according to the question/answer ('v') above, a voltage automatically exists in the space near this charged wire, why do we bother troubling with anything other than this seemingly simple set-up when we wish to operate electrical devices?
Put more specifically:
What about this scenario of One Long Net-Charged Wire does NOT constitute an operable electric circuit?

Describe and discuss the fewest items you would need to add
and/or the fewest things you would need to do

in order to turn this situation
(a long straight wire of uniformly dense net positive charge)

into a simple & functional CIRCUIT.

Your response MUST include at least two complete sentences of English
AND at least one clear and specific picture.

Both the sentences and the picture must be made originally by you (4 pts).

SCORE: ____________

## III. An actual CIRCUIT (15 pts).

Examine the following circuit (values provided directly to its right).

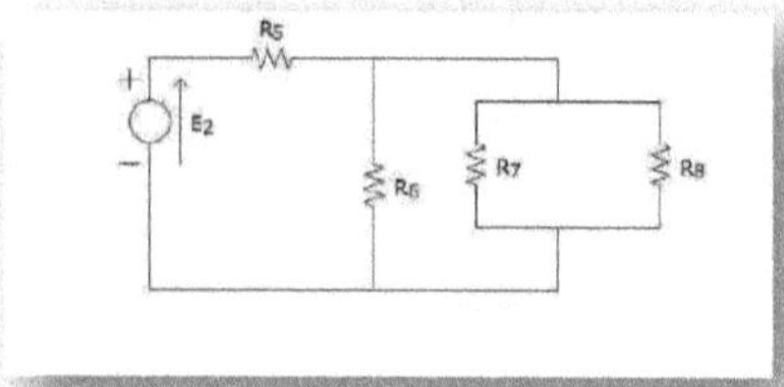


| | |
|---|---|
| E2 | 9 VOLTS |
| R5 | 500 OHMS |
| R6 | 600 OHMS |
| R7 | 700 OHMS |
| R8 | 800 OHMS |

SHOWING ALL WORK, Determine:

a) The ***current*** flowing through each and every RESISTOR (3 pts (i), 2 pts (ii), 1 pt each of (iii) and (iv)).

i. ***I*** (R5) =

ii. ***I*** (R6) =

iii. ***I*** (R7) =

iv. ***I*** (R8) =

b) The ***potential difference*** across each and every RESISTOR (2 pts each).

i. **ΔV** (R5) =

ii. **ΔV** (R6) =

iii. **ΔV** (R7) =

iv. **ΔV** (R8) =

SCORE: ____________

## IV. B-Fields (20 pts).

Be brief but complete and precise regarding all of the following:

A. According to Ampere's historic experimental finding, two light, straight, long wires carrying currents going in parallel directions will do WHAT to each other (1 pts)?

B. Ampere's experimental finding is generally explained by a belief that moving charges create WHAT (1 pts)?

C. Draw a magnetic field line diagram for a long straight current:

a) Head-On: As though the current is coming out of the whiteboard toward your eye (1 pts).

b) Side-View: As though the current is traveling in a straight line from one side of the whiteboard toward the other (1 pts).

D. Concisely explain the essential differences between the "Dot Product" and the "Cross Product" for the multiplication of two vectors (3 pts).

E. Write down a clear and complete expression (EQUATION!) for the magnetic field as a function of charge, velocity and displacement from the charge (2 pts).

F. Write down clear and complete expression (EQUATION!) for the magnetic field as a function current, length and displacement from the current (1 pts).

G. Write down a clear expression for the magnetic force as a function of current, length, and magnetic field (1 pts).

H. Concisely explain the essential differences between the behavior of electric field lines from charges and the behavior of magnetic field lines from charges (3 pts).

SCORE: ___________

I. The POINT.

a) Draw one long and horizontal current-carrying wire a small amount of space above another identical wire – carrying current in the same direction as the wire below it (1 pt).

b) Referring to your picture, you are going to follow a few specific instructions and, ultimately, provide a thorough physics-based explanation of Oersted's historical observation of two current-carrying wires (and what they apparently do when carefully isolated). . .

REQUIREMENTS/DIRECTIONS

i. The explanation you provide MUST include explicit reference to ***EVERY ONE OF*** your answers provided above (A-H). Make a mark (like an asterisk) at the beginning of the sentence each time you have referred to one of the answers (A-H).

ii. The explanation you provide must be strongly focused on the concept of ***DIRECTION***. You need not be too concerned with the concept of magnitude.

iii. Beginning with a little discussion of the field created by the current above (what does it 'look like'? in what directions do the lines point? etc), proceed step-by-step through discussion of the direction of force exerted by this field and, ultimately, how this force affects the current below.

iv. Supplement your initial drawing with any and every visual detail or side-image that might help clarify your explanation.

The goal of all your descriptions is ultimately to make sense out of Oersted's finding in terms of the laws of physics: specifically, why things 'happen' in the direction that they do.
Given your picture and the laws of physics,
your explanation must ultimately answer this overall question:

Under highly controlled conditions,

a horizontal current
will be observed to
accelerate up
if a an identically directed current
is located above it

WHY?!

(5 pts).

SCORE: ___________

## V. *Light: The Excluded Middle (20 pts).*

True/False: Put a "T" in the box next to each claim that appears more true than false.

Put an "F" in the box next to each claim that appears more false than true. (1 pt each).

| | |
|---|---|
| 1. As long as a wave's medium does not change in any respect, then the speed of that wave will not change. | |
| | XXXXXX |
| 2. According to Faraday's Law, if the magnetic flux through some open area does not remain constant in time, then an electric potential will be induced in the closed path bounding that area. | |
| | XXXXXX |
| 3. ***Nearby but outside*** a charging capacitor ($q > 0$), the magnetic field has a magnitude of 0. | |
| | XXXXXX |
| 4. ***Inside*** a charging capacitor ($q > 0$), 0 Coulombs per second of electric charge flows from one plate to the other. | |
| | XXXXXX |
| 5. In a circuit containing a capacitor, no current can flow through the wires until the capacitor is fully charged. | |
| | XXXXXX |
| 6. Given an n-dimensional region of space, the region's boundary is a closed region of dimension n-1. | |
| | XXXXXX |
| 7. A closed surface defines a unique volume. | |
| | XXXXXX |

| | |
|---|---|
| 8. A closed path defines a unique area. | |
| | XXXXXX |
| 9. If an Amperian loop is drawn outside a charging capacitor, then the area bound by the loop must lie outside the capacitor. | |
| | XXXXXX |
| 10. Inside a charging capacitor, 0 magnitude of electric flux flows from one plate to the other. | |
| | XXXXXX |
| 11. Maxwell's Displacement Current correction was added to Gauss's Law and | |

SCORE: ___________

| | |
|---|---|
| allowed Gauss's Law to apply to steady currents. | |
| | |
| 12. According to the explanation of Maxwell's Displacement Current, if the magnitude of an electric flux through an area changes in time, a magnetic field will be induced in the perimeter of that area. | |
| | XXXXXX |
| 13. Mutually induced electric and magnetic fields must be perpendicular to each other. | |
| | XXXXXX |
| 14. Given the function $I(t) = \frac{1}{6}at^3$, no matter how many times you differentiate with respect to t ($\frac{dI}{dt}, \frac{d^2I}{dt^2}, \frac{d^3I}{dI^3}, \ldots$), you will never arrive at the answer "0". | |
| | XXXXXX |
| 15. Given the function $I(t) = \frac{1}{6}e^{3t}$, no matter how many times you differentiate with respect to t ($\frac{dI}{dt}, \frac{d^2I}{dt^2}, \frac{d^3I}{dI^3}, \ldots$), you will never arrive at the answer "0". | |
| | XXXXXX |
| 16. If you find charge in some lab, you will necessarily find electric and/or magnetic field in that lab. | |
| | XXXXXX |
| 17. If you find electric and/or magnetic field in some lab, you will necessarily find charge in that lab. | |
| | XXXXXX |
| 18. When electric and magnetic fields mutually induce each other in a perpetual pattern, that pattern satisfies the wave equation. | |
| | XXXXXX |
| 19. $\frac{1}{\sqrt{\varepsilon_0\mu_0}}$ = the speed at which mutually inducing electric and magnetic fields propagate through a vacuum. | |
| | XXXXXX |
| 20. $\frac{1}{\sqrt{\varepsilon_0\mu_0}}$ = the speed at which light travels through a vacuum. | |
| | XXXXXX |
| 21. *** BONUS *** You will live long and prosper; the net force will be with you. | |

SCORE: ___________

2

# FULL-TEXT OF FINAL EXAM SU15: SOLVED & EXPLAINED

# FINAL EXAM: PHYSICS 204, JULY 17, 2015.

~,Charge, Flow, Field, Flux & Lu},.

JOHN JAY COLLEGE OF CRIMINAL JUSTICE,
THE CITY UNIVERSITY OF NEW YORK

DANIEL A. MARTENS YAVERBAUM

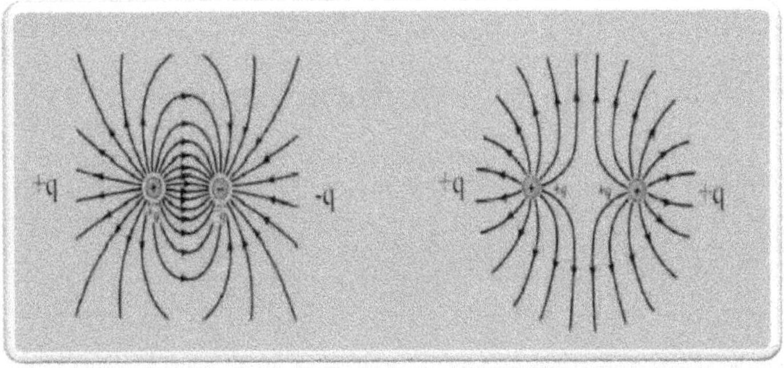


**Name:** SOLUTIONS

**Section #:** ____________

SCORE: ____________

# SOME USEFUL RELATIONS:

1) $\oint \vec{E} \cdot \overrightarrow{da} = \frac{q_{(enc)}}{\varepsilon_0}$.

2) $\oint \vec{B} \cdot \overrightarrow{da} = 0$.

3) $\oint \vec{E} \cdot \overrightarrow{dl} = -\frac{d}{dt} \int \vec{B} \cdot \overrightarrow{da}$.

4) $\oint \vec{B} \cdot \overrightarrow{dl} = \mu_0 I_{(enc)} + \mu_0 \varepsilon_0 \frac{d}{dt} \int \vec{E} \cdot \overrightarrow{da}$.

5) $\vec{E} = \frac{1}{4\pi\varepsilon_0} \frac{q}{r^2} \hat{r}$.

6) $\vec{F}_E \equiv q\vec{E}$.

7) $\vec{B} = \frac{\mu_0}{4\pi} \frac{q}{r^2} \vec{v} \times \hat{r}$.

8) $\vec{F}_B \equiv q\vec{v} \times \vec{B} = \vec{I}l \times \vec{B}$.

9) $V_a - V_b \equiv \int_a^b \vec{E} \cdot \overrightarrow{dr}$.

10) $C \equiv \frac{Q}{\Delta V}$. (Note: This "C" is capital.)

11) $I \equiv \frac{dq}{dt}$

12) $I = \frac{\Delta V}{R}$.

13) $\mathcal{E} - IR - \frac{Q}{C} = 0$. (Capital "C".)

14) $c \approx 3 \; x \; 10^8 \; m/s$.

15) $n \equiv \frac{c}{v}$. (Lower-case "c".)

16) $n_1 \sin\theta_1 = n_2 \sin\theta_2$.

17) $\frac{Yd}{L} = \frac{n\lambda}{2}$

18) $sin^2\theta + cos^2\theta = 1.$

19) $\bar{v} \equiv \frac{\Delta x}{\Delta t}.$

20) $\sum \vec{F} = m\vec{a}.$

21) $F = -Kx.$

22) $x = Acos(\omega t + \phi).$

23) $\omega = 2\pi f.$

24) $f = \frac{1}{T}.$

25) $KE = \frac{1}{2}mv^2$

26) $PE_{elastic} = \frac{1}{2}Kx^2$

27) $v = \lambda f.$

28) $v = \sqrt{\frac{T}{\mu}}.$

29) $\frac{\partial^2 y}{\partial t^2} = (v^2)\frac{\partial^2 y}{\partial x^2}.$

30) $\epsilon_0 \approx 8.85 \; x \; 10^{-12} \; \frac{C^2}{Nm^2}.$

31) $\mu_0 \approx 1.26 \times 10^{-7} \; \frac{N}{A^2}.$

32) $m_e \approx 9.11 \; x \; 10^{-31}$ kg.

33) $m_{(PROTON)} \approx m_{(NEUTRON)} \approx 1.67 \; x \; 10^{-27}$ kg.

34) $q_e \approx 1.60 \; x \; 10^{-19} \; C.$

## I. E-FIELDS FROM ***POINT CHARGES*** (20 PTS).

Two ***point-charges*** of differing magnitudes and are held stationary in an enormously spacious x-y plane.

A researcher places an instrument called a 'field detector' at the point (-5,+12). She is interested in measuring the electric field at that precise location.

The two point charges are as follows:

| Name | Charge | x-Coordinate | y-Coordinate | Ordered Pair |
|---|---|---|---|---|
| $Q_1$ | $+5 \times 10^{-10}$ Coulombs | +5 meters | 0 meters | (5,0) |
| $Q_2$ | $-12 \times 10^{-10}$ Coulombs | 0 meters | -12 meters | (0,-12) |

**Location of Interest:** (-5,+12)

Note: All coordinates are measured and given in **meters** (not centimeters); similarly, the (enormous) charge magnitudes are in whole **Coulombs** (not micro-Coulombs).

Also Note: If you wish, you are permitted and encouraged to approximate the electrostatic constant as:

$$K_e \approx 10 \times 10^9 \approx 1 \times 10^{10} \frac{Nm^2}{C^2}.$$

* * *

a) Draw a neat and clear sketch of the situation, as you understand it. Your sketch must express a clear decision as to which directions are designated by + and – on each axis (3 pts).

PLEASE SEE NEXT PAGE . . .

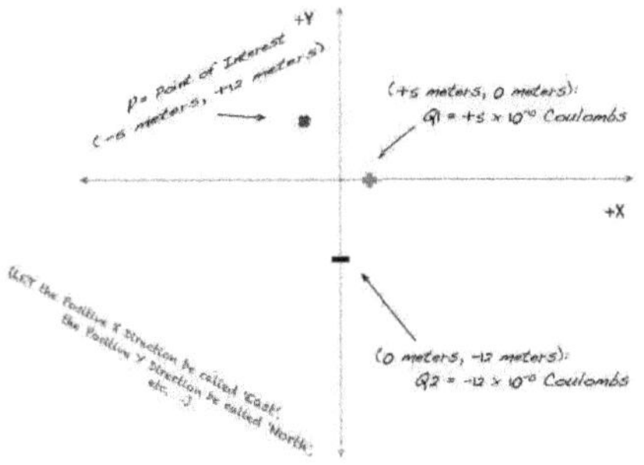


b) For *each* of the two individual point charges, draw an approximate field line diagram – each drawn as it would look IF the other charge did NOT exist. For BOTH, however, obey the following convention: For every $1 \times 10^{-10}$ Coulombs of strength that generates a field, at least one more field line should appear in the field line diagram (3 pts).

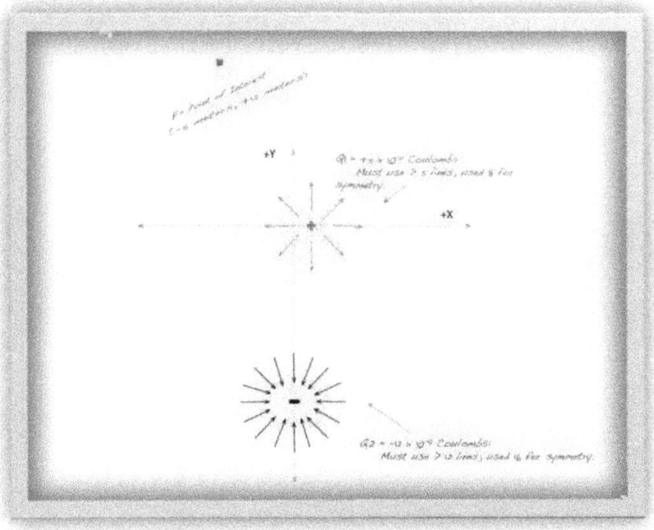

c) Compute the Electric Field as measured at the Point of Interest (-5,+12).

That is:

i. In Newtons/Coulomb, determine the ***electrostatic field*** <u>**magnitude**</u> at this location of interest (5 pts).

X --> ...

$$\sum E_x = K_e\left[\left(\frac{Q_1}{r_1^2}\cdot\frac{x_1}{r_1}\right)+\left(\frac{Q_2}{r_2^2}\cdot\frac{x_2}{r_2}\right)\right]$$

$$\sum E_x = K_e\left[\left(\frac{5\times10^{-10}}{244}\cdot\frac{-10}{\sqrt{244}}\right)+\left(\frac{12\times10^{-10}}{601}\cdot\frac{5}{\sqrt{601}}\right)\right]$$

$$\sum E_x = K_e(1\times10^{-10})\left[\left(\frac{5}{244}\cdot\frac{-10}{\sqrt{244}}\right)+\left(\frac{12}{601}\cdot\frac{5}{\sqrt{601}}\right)\right]$$

$$\sum E_x \approx \cancel{(1\times10^{10})}\,\cancel{(1\times10^{-10})}\left[\left(\frac{5}{244}\cdot\frac{-10}{\sqrt{244}}\right)+\left(\frac{12}{601}\cdot\frac{5}{\sqrt{601}}\right)\right]$$

$$\sum E_x \approx \left[\left(\frac{-50}{3.81\times10^{3}}\right)+\left(\frac{60}{1.47\times10^{4}}\right)\right]$$

$$\approx\left[(-1.31\times10^{-2})+(4.08\times10^{-3})\right]$$

$$\sum E_x \approx -9.02\times10^{-3}\,N/C$$

Y --> ...

$$\sum E_y = K_e\left[\left(\frac{Q_1}{r_1^2}\cdot\frac{y_1}{r_1}\right)+\left(\frac{Q_2}{r_2^2}\cdot\frac{y_2}{r_2}\right)\right]$$

$$\sum E_y = \cancel{K_e}/\cancel{K_e}\left[\left(\frac{5}{244}\cdot\frac{12}{\sqrt{244}}\right)+\left(\frac{12}{601}\cdot\frac{-24}{\sqrt{601}}\right)\right]$$

$$\sum E_y \approx \left[\left(\frac{60}{3.81\times10^{3}}\right)-\left(\frac{288}{1.47\times10^{4}}\right)\right]$$

$$\sum E_y \approx \left[(1.57\times10^{-2})-(1.96\times10^{-2})\right]$$

$$\sum E_y \approx \left[(1.57\times10^{-2})-(1.96\times10^{-2})\right]$$

$$\sum E_y \approx -3.90\times10^{-3}\,N/C$$

$$\vec{\mathbf{E}} = \sum E_x + \sum E_y$$

Continued on next page . . . →

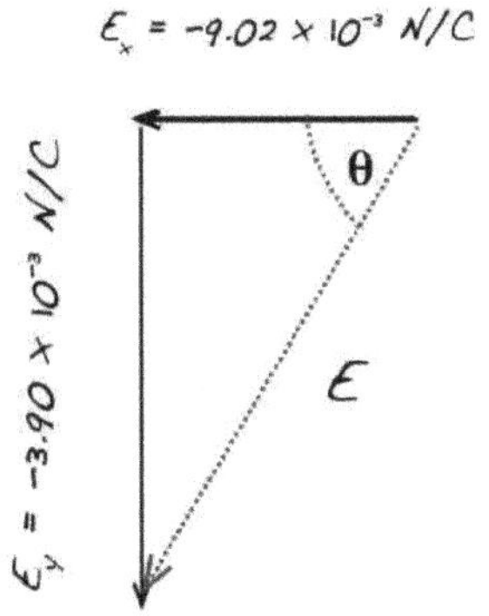


$$\vec{\mathbf{E}} \equiv \sum E_x + \sum E_y$$

$$E \equiv \|\vec{\mathbf{E}}\| \approx \sqrt{(9.02 \times 10^{-3})^2 + (-3.90 \times 10^{-3})^2}$$

$$\boxed{E \approx 9.83 \times 10^{-3}\, N/C}$$

ii. Using degrees **and** points of the compass (such as "20 degrees South-West"), express the **direction** of the ***electrostatic field*** at this Point of Interest (-5,+12) (3 pts).

$$\theta \approx \tan^{-1}\left(\frac{3.90 \times 10^{-3}}{9.02 \times 10^{-3}}\right)$$

$$\theta \approx 1.03\, radians\ NW\ \ OR\ldots$$

$$\boxed{\theta \approx \ \ 23.4^\circ\ South - West}$$

*So,*

$$\boxed{\vec{E} \approx 9.83 \times 10^{-3}\, N/C\ at\ 23.4^\circ\ South - West}$$

d) Staying approximately consistent with whatever scale, style (etc.) was involved in your diagrams for (b), above, try now to bring the two graphics together into one visual ***superposition*** of the situation: That is, draw one field line diagram for the net influence exerted by this pair of point charges. In your mind, start bringing the two pictures increasingly close together... Since lines cannot, however, cross nor disappear, allow denser regions of lines from one charge to gently and smoothly 'bend back' (or forward) the lines from another – doing as much as you can to preserve simplicity and symmetry where applicable (3 pts).

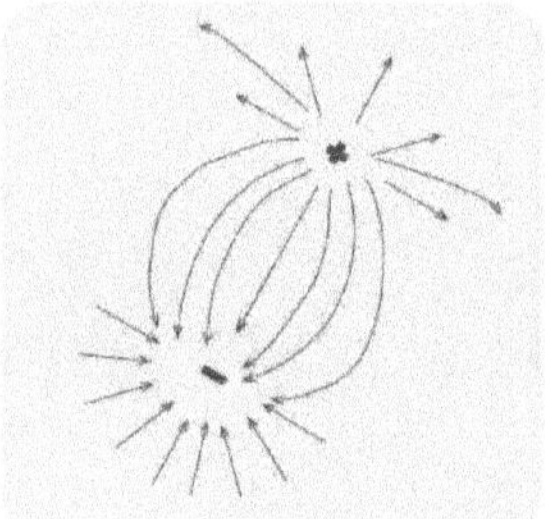

e) Assume that a somehow isolated and highly condensed Helium Nucleus (consisting two protons, two neutrons and no other measurably significant entities) is introduced and held gently at the point (-5,12). Assume, further, that any uncharged particles trapped in that nucleus simply 'go along for the ride' whenever the charged particles are motivated to travel. The nucleus is then released. Compute the nucleus 's initial instantaneous acceleration. Provide precise **magnitude** AND **direction** (3 pts).

PLEASE SEE NEXT PAGE

for solution sample . . .

$\vec{\mathbf{F}}_e \equiv q\vec{\mathbf{E}}.$

$$Here, q = 2\ protons.$$
$$\vec{\mathbf{F}}_e \approx (2 \times 1.60 \times 10^{-19} C)\vec{\mathbf{E}}$$
$$\vec{\mathbf{F}}_e \approx (3.2 \times 10^{-19} C)(9.83 \times 10^{-3} N/C)$$
$$\vec{\mathbf{F}}_e \approx 3.15 \times 10^{-21} N \text{ at} \sim 23.4^\circ \text{ South} - \text{West}$$

$$\sum \vec{\mathbf{F}} = m\vec{a}.$$

*Assuming that this electrostatic field is isolated from all other influences, then the force exerted by the field is the only one acting on the helium nucleus...*

$$\vec{\mathbf{F}}_e = m\vec{a}$$
$$a = \| \vec{a} \| = \frac{F_e}{m_{hn}}$$
$$m_{hn} \equiv mass\ of\ helium\ nucleus \approx 2p + 2n \approx 4p$$
$$m_{hn} \approx 4(1.67 \times 10^{-27} kg)$$
$$a \approx \frac{(3.15 \times 10^{-21} N)}{(6.68 \times 10^{-27} kg)}$$

*At a given point, any positive charge is accelerated in the SAME direction as the (tangent line to the) field line at that point:*

$$\boxed{a \approx 4.71 \times 10^5 m/s^2 \text{ at} \sim 23^\circ \text{ S} - \text{W}.}$$

## II. Gauss's Law (25 pts).

A huge but imaginary sphere is drawn with its center at the origin of a coordinate system, such as that used in Problem I, above.
The radius of the sphere is ***r* = 15 meters**.

**Two point charges have been sitting within a 15 meter radius of the origin; they continue to sit there.**

One charge has a magnitude of +5 x 10-10 Coulombs;
the other charge has a magnitude of -12 x 10-10 Coulombs.
Nothing else exists within 15 meters of the origin.

a) *Draw this situation* as you understand it (2 pts).

(This is not meant to be a trick; the drawing will just help clarify.)

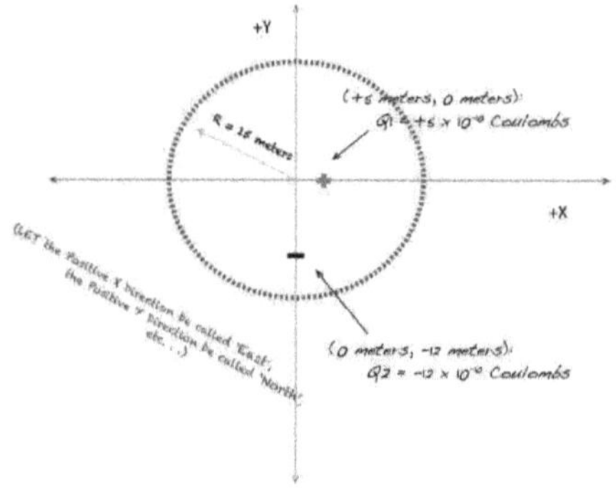

b) In $\frac{Newtons \cdot Meters^2}{Coulombs}$, find the total amount of ***electric flux*** that passes through the surface of this imaginary sphere (2 pts).

Your answer should be a number, expressed in the units mentioned directly above.

*Definition of Electrostatic FLUX:*

$$\Phi_E \equiv \oint \vec{E} \cdot d\vec{A}$$

*Gauss's Law regarding Electrostatic FLUX:*

$$\oint \vec{E} \cdot d\vec{A} = \frac{q_{(enc)}}{\varepsilon_0}$$

(The electrostatic flux through ANY closed surface
is always and simply the total magnitude of the charge
enclosed by that surface - -divided by a constant
(known as the 'permittivity of free space') - -
entirely independent of the size or shape
of the closed surface;
entirely independent of the shape
of the enclosed charge
and
entirely independent of ANY charges located anywhere
OUTSIDE the closed surface.
(!)
So, the answer is simply:

$$\frac{(5.00 \times 10^{-10}\,Coulombs) - (12.0 \times 10^{-10}\,Coulombs)}{\varepsilon_0}$$

$$\approx \frac{-7.00 \times 10^{-10}\,Coulombs}{8.85 \times 10^{-12}\,C^2/_{N\,m^2}}$$

$$\boxed{\Phi_E = q_{(enc)} \approx -79.1\,Nm^2/C!}$$

(i.e.: LINE) that is net-positively charged; c

*Your first large goal, after a series of smaller 'build-up' questions to follow on the next page, will be to compute the electric field as a function of perpendicular distance,* ***r****, from this line of charge...* (continued on next page) . . .

i. Draw the field line diagram produced by this long line of positive charge (2 pts).

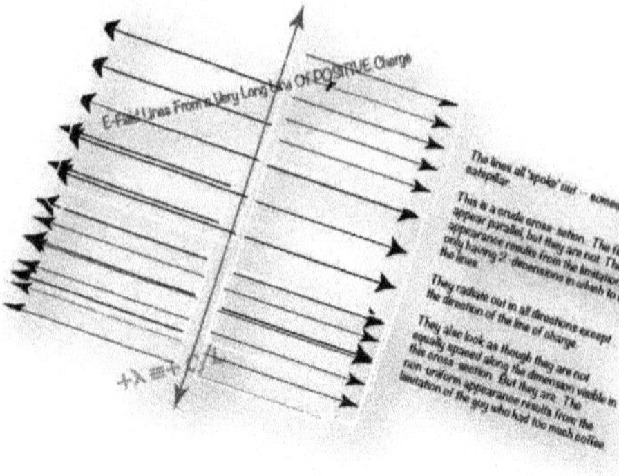


ii. In a sentence or two of English, explain why the sphere drawn above will not be the most convenient shape for computing ***electric field*** at ***r*** (3 pts).

Gauss's Law is true for ANY closed surface, so a sphere CAN absolutely be drawn around a vertically oriented continuous line of charge. The total flux through the surface of that sphere WILL be the total charge ($\sim\lambda L$) enclosed by that sphere - despite the fact that a sphere produces field lines that are symmetric in all directions with NO exception ("0-dimensional symmetry") while a line is symmetric in all directions save ONE (its own axis - hence "1-dimensional symmetry"). Truth is not the problem. The problem is that this difference in pattern (between a line and a sphere) means that the lines spoking from the line of charge will all 'flow' (***flux***) through the sphere surface at different angles and in densities that will differ from region to region. The magnitude and the direction of the field lines, that is, will NOT BE A CONSTANT through the surface of that sphere. So the E-Field term in the Flux Integral CANNOT be 'taken out' of the integral and we actually have to compute an integral. And not an easy one, to say the least. A central purpose of Gauss's Law is to avoid cumbersome or intractable integral calculations by thinking conceptually.... Physically instead. To trade math for physics. If we draw a sphere around the line of charge, Gauss's Law is still right, but it just isn't helpful. We want to set up an integral that we do NOT actually have to do.

iii. Choose and draw a Gaussian (closed) surface surrounding some portion of the line—a surface that will best help you compute the field at $\boldsymbol{r}$ (2 pts).|

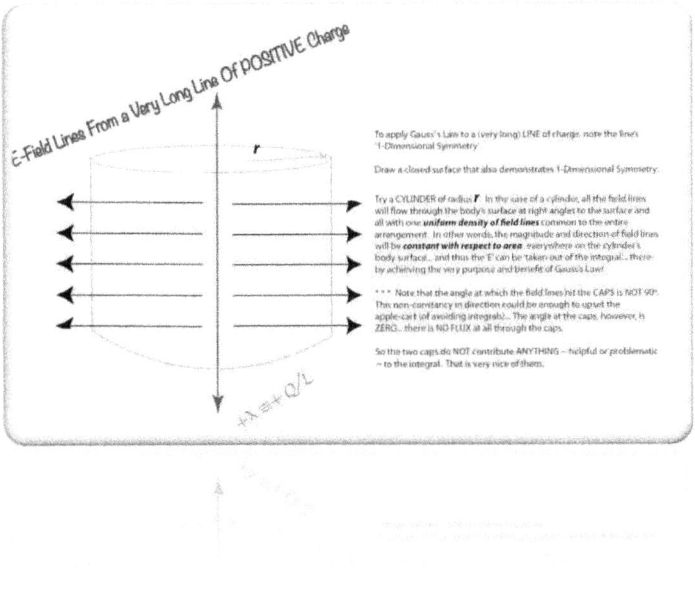
E-Field Lines From a Very Long Line Of POSITIVE Charge
r
+λ ≡ + Q/L
To apply Gauss's Law to a (very long) LINE of charge, note the line's '1-Dimensional Symmetry'.
Draw a closed surface that also demonstrates 1-Dimensional Symmetry.
Try a CYLINDER of radius r. In the case of a cylinder, all the field lines will flow through the body's surface at right angles to the surface and all with one **uniform density of field lines** common to the entire arrangement. In other words, the magnitude and direction of field lines will be **constant with respect to area** everywhere on the cylinder's body surface... and thus the 'E' can be 'taken out of the integral'... thereby achieving the very purpose and benefit of Gauss's Law!
* * * Note that the angle at which the field lines hit the CAPS is NOT 90°. This non-constancy in direction could be enough to upset the apple-cart (of avoiding integrals)... The angle at the caps, however, is ZERO... there is NO FLUX at all through the caps.
So the two caps do NOT contribute ANYTHING – helpful or problematic – to the integral. That is very nice of them.

iv. Starting with Gauss's Law and proceeding through as many clear and verbally explained steps as possible,

find $\boldsymbol{E}$ as a function of (perpendicular distance) $\boldsymbol{r}$
(given $\boldsymbol{\lambda}, \boldsymbol{\varepsilon_0}$)
from this line of charge (5 pts).

This answer will not be a number.

$$\oint \vec{\mathbf{E}} \cdot d\vec{\mathbf{A}} = \frac{q_{(enc)}}{\varepsilon_0}$$

$$E \oint dA = \frac{q_{(enc)}}{\varepsilon_0}$$

(See explanation in 2 (b) ii, above.)

NOTE: On the one hand, if you cannot legitimately make this simplification, then Gauss's Law does you no good.

Applied properly, the E will ALWAYS 'come out' of the integral.

On the other hand, you cannot simply skip over all the drawing and over the CHOOSING of the most convenient closed ('Gaussian') surface to draw:

It might well seem as though all the drawing and choosing is all nonsense - -

that we should just write down $EA = \frac{q(enc)}{\varepsilon_0}$ and get on with it

since it's apparently going to be true every time.

*. . . But Gauss was no fool. Neither be you! Should you wish to drain the bathwaters of thick integral computation, then clutch tightly the babies of well-chosen surface and meaningful integral set-up! . . .* →

On the contrary: The drawing and choosing is the heart of the method.

Without drawing and choosing, you don't know WHAT SURFACE AREA is being used

(supposedly by you!) and thus you DO NOT KNOW by what expression you will divide FLUX

in order to get FIELD - -

Determination of the FIELD WAS AND IS THE GOAL of considering Flux. So... back to this calculation

of the E - Field at a distance r from a very long line of charge, . . .

$$\ldots E \oint dA = \frac{q_{(enc)}}{\varepsilon_0}$$

$$EA = \frac{q_{(enc)}}{\varepsilon_0}$$

In this particular example/application, therefore,

'area' refers to the surface area of a CYLINDER (radius r, length L):

Surface Area(cylinder) = $2\pi rL$.

$$E \cdot 2\pi rL = \frac{q_{(enc)}}{\varepsilon_0}$$

$$E = \frac{1}{2\pi\varepsilon_0}\frac{q}{L}\frac{1}{r} \ldots \textit{by definition, then,} \ldots$$

$$\boxed{E = \frac{1}{2\pi\varepsilon_0}\frac{\lambda}{r}!}$$

v. Now assume that $\lambda \equiv 5\frac{Coulombs}{meter}$.

In VOLTS, Find the ***Electric Potential Difference*** ('Voltage'!) that exists (as long as that line remains net-charged) between the point **(5,-6)** and the point **(10,-18)**.

This is not a joke. Nor a trick.
Hint: Recall and deploy the definition of ***potential difference*** (5 pts).

See next page for solution . . .

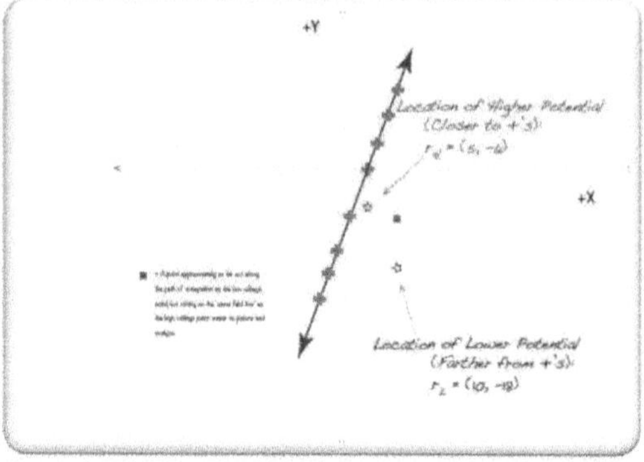


Consider the meaning of potential difference:

$$\Delta V \equiv V_{(\text{at a place of higher potential})} - V_{(\text{at a place of lower potential})} \equiv V_H - V_L$$

$$V_H - V_L \equiv \int_{H}^{L} \vec{\mathbf{E}} \cdot \vec{dr}$$

Each Volt of Potential is a Joule per Coulomb:

The potential difference between two locations is
the work 'per charge' that the field will do
to restore each + 1 Coulumb 'test charge'
to a more stable location
(to bring each test charge closer, that is, back to 'ground').

Field lines, given this definition, always point
from places of Higher Potential to places of Lower Potential:

Locations closer to net positive charge distributions are places of higher potential.

In this case, we will integrate from the 'High Voltage' point at (5, - 6) to the 'Low Voltage' point at (10, - 18).

$$\int_H^L \vec{\mathbf{E}} \cdot \overrightarrow{dr} = \int_{H=P_H(5,-6)}^{L=P_L(10,-18)} \vec{\mathbf{E}} \cdot \overrightarrow{dr}.$$

Note : On Power Supplies, Batteries and other commercial devices,
that's what the '+' and '−' mean:
locations of comparatively higher and lower electrostatic potential.

But the limits of integration are icing on the cake;
the integration itself is the more fundamental issue and worth more points in an exam.
Let's look at it first.

$$\Delta V = \int_H^L \vec{\mathbf{E}} \cdot \overrightarrow{dr}$$

What is $\vec{E}$? That was the whole purpose of deploying Gauss's Law:
To find $\vec{E}$ AS A FUNCTION of r.

From II (c) (iv), above, we have:

$E = \frac{1}{2\pi\varepsilon_0}\frac{\lambda}{r}$. So, here:

$$\Delta V = \int_H^L \vec{\mathbf{E}} \cdot \overrightarrow{dr} = \int_H^L \frac{1}{2\pi\varepsilon_0}\frac{\lambda}{r} dr$$

$$\Delta V = \frac{\lambda}{2\pi\varepsilon_0}\int_H^L \frac{1}{r} dr \quad . \; . \; .$$

$$\Delta V = \left(\frac{\lambda}{2\pi\varepsilon_0}\right)\ln r \Big]_{r_H}^{r_L}.$$

continued on next page . . .

The field function has been integrated.

All constants are known.

($\lambda$ has been given as 5 C/m.)

It's now time to evaluate-

so as to find an actual numerical answer (in Volts).

## Remember:

$$\Delta V = \int_{r\,at\,P(5,-6)}^{r\,at\,P(10,-18)} \vec{\mathbf{E}} \cdot \overrightarrow{dr}$$

Just like calculating Work,

$r_{\parallel}$ ≡ Displacement (from Charge Line. to Point of Interest), BUT

ONLY the Component PARALLEL to Field Line (direction of force).

How should we think about those limits?

They might well look confusing,

but only because our priorities have shifted a bit since the beginning of the problem.

The points were presented in a manner most consistent

with the rest of the problem:

as coordinate pairs on a (2-D) Cartesian plane.

Now that electric potential has been introduced, however,

we want to picture the locations as endpoints on some (1-D) line-- of integration.

One of many general ways to see/calculate the distance between a line and a point is below.
You can get this distance whatever way you like;
you can even approximate it without penalty.
The important thing is to recognize that
'distance from a line' necessarily and exclusively refers to paths perpendicular to that line.
This is mathematically true by definition,
and it is emphatically true in this physics context:
We are summing DOT PRODUCTS between field and the displacement.
The field lines are all perpendicular to the line of charge.

In general, the distance between a line and a point =

$$\frac{ax_0 + by_0 + c}{\sqrt{a^2 + b^2}}.$$

Our charge line is:

$$y = \frac{12}{5}x - 12.$$

The expression of our line most convenient for this context is:

$$\frac{12}{5}x - 1y - 12 = 0.$$

*That is,*

$$12x - 5y - 60 = 0.$$

*So,*

$$r_H = \frac{12x_0 - 5y_0 - 60}{\sqrt{25 + 144}} = \frac{12x_0 - 5y_0 - 60}{13}$$

$$r_H = \frac{12(5) - 5(-6) - 60}{13} = \frac{\cancel{60} + 30 - \cancel{60}}{13}$$

$$r_H \approx 2.31\,m\,.$$

continued on next page. . .. . . →

r($L$) can be found similarly (from same line with, therefore, same equation):

$$r_L = \frac{12x_0 - 5y_0 - 60}{13}$$
$$r_L = \frac{12(10) - 5(-18) - 60}{13} = \frac{120 + 90 - 60}{13} = \frac{150}{13}$$
$$r_L \approx 11.5$$

*SO*:

$$\Delta V = \int_{r_H \approx 2.31}^{r_L \approx 11.5} \vec{\mathbf{E}} \cdot \vec{dr}$$

$$\Delta V = \left(\frac{\lambda}{2\pi\varepsilon_0}\right) \ln r \Big]_H^L$$

$$\Delta V = \left(\frac{\lambda}{2\pi\varepsilon_0}\right) \ln r \Big]_{r_H \approx 2.31}^{r_L \approx 11.5}$$
$$\Delta V = \left(\frac{\lambda}{2\pi\varepsilon_0}\right) [\ln(r_L) - \ln(r_H)]$$

$$here, \lambda = 5\, C/m.$$

$$\Delta V \approx \left(\frac{5}{2\pi\varepsilon_0}\right) \ln r \Big]_{r_H \approx 2.31}^{r_L \approx 11.5}$$

$$\Delta V \approx \left(10 \times \frac{1}{4\pi\varepsilon_0}\right) \ln r \Big]_{r_H \approx 2.31}^{r_L \approx 11.5} \approx (10 \times K_e) \ln r \Big]_{r_H \approx 2.31}^{r_L \approx 11.5}$$
$$\Delta V \approx (1 \times 10^{11}) \ln r \Big]_{r_H \approx 2.31}^{r_L \approx 11.5}$$
$$\Delta V \approx (1 \times 10^{11}) \ln r \Big]_{r_H \approx 2.31}^{r_L \approx 11.5}$$
$$\Delta V \approx (1 \times 10^{11})[\ln(11.5) - ln(2.31)]$$
$$\Delta V \approx (1 \times 10^{11})[2.44 - .837]$$

$$\boxed{\Delta V \approx 1.60 \times 10^{11}\, Volts}$$

Note: This amount of 'voltage' is preposterously high.
But so is 5 Coulombs of charge for each of the preposterously large meters we investigate in this problem.
And the number 5 is easier to work with than, say, some other numbers.
As always, we have to make choices:
Chalk them up to 'conservation of convenience.'

Also NOTE: This question would have been slightly easier to picture and approach (though no less challenging to solve) had the second given point looked more like something such as, for example, (12, -5). The second given point was miscalculated, so did not look particularly encouraging.

The problem is the same and solvable by the same method either way. The question, however, is likely to seem clearer and more approachable if the two given points lie along a line manifestly perpendicular to the original line of charge. In such case, the relevant displacement (and therefore path of integration) is directly from one point to the other. Such was the intention of the question, but the second point was miscalculated. As the question is presented, we still solve precisely the same way – 'walking' and integrating along a field line directly away from the charge line, ignoring all other components of displacement, from the first point of interest to the second. When we do not end up 'standing' on the second point of interest, it is undoubtedly harder to see, remember and believe that we need only count the component of displacement which is parallel to the field line. It is nonetheless true either way.

vi. Batteries and power supplies are generally used to supply ***Electric Potential Difference*** ('Voltage') to circuits, thereby allowing us to operate electric and electronic devices. If, according to the question/answer ('v') above, a voltage automatically exists in the space near this charged wire, why do we bother troubling with anything other than this seemingly simple set-up when we wish to operate electrical devices?

Put more specifically:

What about this scenario of One Long Net-Charged Wire does NOT constitute an operable electric circuit?

**Describe and discuss the fewest items you would need to add and/or the fewest things you would need to do**

**in order to turn this situation (a long straight wire of uniformly dense net positive charge)**

**into a simple & functional CIRCUIT.**

Your response MUST include at least two complete sentences of English AND at least one clear and specific picture.

Both the sentences and the picture must be made originally by you (4 pts).

*This situation gives charges a reason to accelerate, but it does not provide a (conductive) path through which they (electrons) are free to do so. Furthermore, we have no reason to be confident that the potential difference provided by a line of charge will remain available or at predictable values once charges begin to flow—particularly if the charges come from and thus deplete the line itself. Finally, we have nothing in place to insure or safely harness the conservation of energy: Were charges somehow to begin flowing in a closed loop, the flow needs to be tempered by some significantly non-conductive material which will decrease their electric potential energy - perhaps converting it to a thermal or otherwise useful form. . .*

Note: In other words, we do indeed have a potential difference between two spots, but that's more likely to be true than not in any arbitrary region of space and time. What we need – and what human civilization did not begin to develop as a complete package until the nineteenth century – are:

1) a way to maintain the particular charge imbalance underlying a potential difference even as charges themselves begin to move: This is the heart of a *Voltaic Cell* and therefore of a *Battery* (or *Power Supply*);
2) a closed loop made of largely conductive material: *Wire*;
3) at least one finite and deliberately placed region with a known extent of low conductivity: *Resistance*. (The path that allows for flow could be the same as the path that hinders flow (it could simply be somewhere in the middle of the conducting-insulating spectrum of materials), but then it would be very difficult to control, modify or analyze.)

So, . . . Make sure that the length of the line of + charge is indeed way beyond the dimensions of everything else we add: Insure that the λ can function as a 'reservoir' of net charge. Connect a strand of wire from the reservoir to one end of a resistor. Connect the other end of the resistor to $r_L$ or, better, to some equivalently huge yet uncharged and reasonably conductive nearby material – such as planet Earth. Allow the huge and neutral object to act as a reservoir for receiving net +'s without disturbance and rely on it to remain $r_L$ no matter how intricately things are added to the circuit, i.e. to serve as the 'Ground' for all conceivable potential energy comparisons.

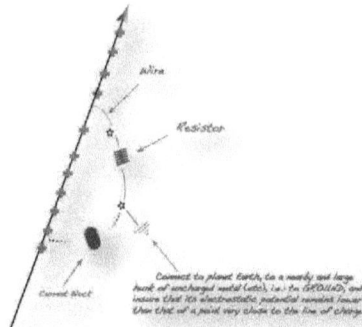

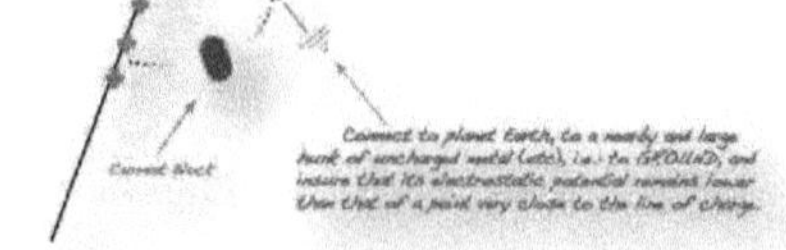

Wire
Resistor
Connect to planet Earth, to a nearby and large hunk of uncharged metal (etc.), i.e. to GROUND, and insure that its electrostatic potential remains lower than that of a point very close to the line of charge.

## III. An actual CIRCUIT (15 pts).

Examine the following circuit (values provided directly to its right).

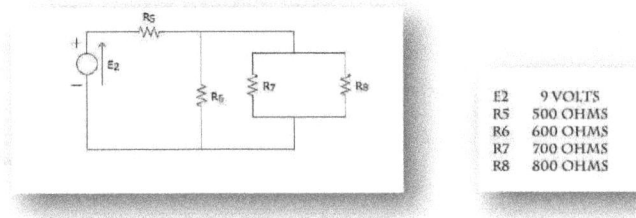

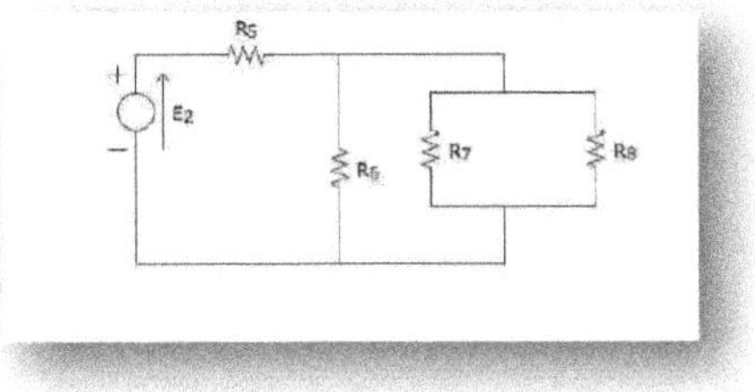


| | |
|---|---|
| E2 | 9 VOLTS |
| R5 | 500 OHMS |
| R6 | 600 OHMS |
| R7 | 700 OHMS |
| R8 | 800 OHMS |

SHOWING ALL WORK, Determine:

a) The ***current*** flowing through each and every RESISTOR (3 pts (i), 2 pts (ii), 1 pt each of (iii) and (iv)).

i. ***I*** (R5) =

$$\frac{1}{R_{eq(R5,R7,R8)}} = \frac{1}{600} + \frac{1}{700} + \frac{1}{800}$$

$$\frac{1}{R_{eq(R2,R3,R4)}} \approx 4.34 \times 10^{-3}\, Ohms^{-1}$$

$$R_{eq(R2,R3,R4)} \approx 230\, Ohms$$

*Therefore,*

$$R_{eq(R1,R2,R3,R4)} \approx 500\, Ohms + 230\, Ohms$$

$$R_{eq(R1,R2,R3,R4)} \approx 730\, \Omega$$

$$I_{Battery} \equiv I_{MainLoop} \equiv \frac{\mathcal{E}}{R_{eq}}$$

*Here,* $I_5 = I_{Battery}$;

$$I_5 \approx \frac{9\, Volts}{730\, Ohms} \approx 1.23 \times 10^{-1} Amperes$$

$$\boxed{I_5 \approx 12.3\, milliAmperes}$$

ii. $\boldsymbol{I}$(R6) = . . .

Now, between ANY TWO POINTS in a circuit,

$$I = \frac{\Delta V}{R}.$$

A central implication of this relation (Ohm's Law) is this:

$$I \propto \frac{1}{R}$$

Given a choice between two current branches,
the portion (fraction) of current that 'chooses' one particular branch is necessarily
INVERSELY proportional to the the fraction of total resistance found in that branch.

Mathematically, treatment of an inverse proportionality might seem 'obvious', but remember:
... You can rely on the seemingly simple pattern 'one goes up, the other goes down'
IFF the third term of the equation – – the one here referring to potential drop – – is a CONSTANT.
$\Delta V$ must be one single value applicable to both branches, NOT, for example,
some kind of 'total value that gets split' between the two!

Indeed for any two paths or devices in parallel, the potential difference
is necessarily a constant . But this somewhat surprising (hard to remember) idea
comes from physics, not from math:

EVERY CHARGED PARTICLE - no matter which way it goes -
and therefore EVERY INDEPENDENT PATH that any charge might possibly follow
is independently subject to the laws of physics.
EVERY POSSSIBLE PATH must therefore INDEPENDENTLY
OBEY ENERGY CONSERVATION.

So, any and all ways to get from point A to point B
within the same circuit
must do the same work
and impose the same energy effect
on each Coulomb of charge:

Therefore,
The potential drops across any two 'parallel' (independent) routes through a circuit
Are necessarily identical.

It is crucial that you are comfortable with the reasoning behind this certainty – –
More often than not, a person who feels hopelessly stuck
in the middle of a circuit problem is forgetting this vital piece of information.
And more often than not, s/he is forgetting it
because s/he doesn't really believe it. . .

(continued on next page).

So, . . .

$I \alpha \frac{1}{R}$

This ~12.3 milliAmperes of current splits into three different portions,

but, much like considering forces when applying Newton's 2nd Law,
it is far clearer and more constructive to think of things
'like a charge': ONE CHOICE AT A TIME.
That is, each charge FIRST faces a split between Way #1 (through R6) or the remaining set of Ways: #2&#3, . . .
THEN a split between Way #2 (through R7) or Way #3 (through R8)

In other words, there really is no meaning or utility to any notion of a '3–Way Choice'.
That's not a choice. That's a mess.

MUCH like this paragraph
might cease to be... if you are willing to slog through a bit more and ...
continue reading on the next page ...

SO: . . .
R5 (500Ω) is in series with the battery: ALL 12.3 mA come through this resistor.

Then this 'Main Loop' of current (or 'Battery Current')
(two equally acceptable terms for, essentially, the trunk of a tree)
splits into a total of [973] possible Ohms:
a 600 Ω path vs a 373 Ω path
$because \left(\frac{1}{700}+\frac{1}{800}\right)^{-1} \approx 373$.

So, $\frac{373}{973}$ of 12.3 mA flows through the the 600Ω ≈ 4.72 mA
and $\frac{600}{973}$ of 12.3 mA flows through the remaining split (700 Ω / 800 Ω) ≈ 7.58mA.

Then, between that 700 Ω and the 800 Ω (a total of 1500 Ω):

$\frac{800}{1500}$ of mA flows through the 700 Ohm ≈ 4.04 mA
and $\frac{700}{1500}$ of 2.41 mA flows through the 800 Ohm ≈ 3.54 mA.

*In conclusion:*

| |
|---|
| *i.* $R_5(500\,\Omega)$: $I_5 \approx 12.3\ mA$. |
| *ii.* $R_6(600\Omega)$: $I_6 \approx 4.72\ mA$. |
| *iii.* $R_7(700\Omega)$: $I_7 \approx 4.04\ mA$. |
| *iv.* $R_8(800\Omega)$: $I_8 \approx 3.54\ mA$. |

b) The ***potential difference*** across each and every RESISTOR (2 pts each).

i. $\boldsymbol{\Delta V}$(R5) =

$\Delta V = IR.$

*So...*

$\Delta V_5 = I_5 R_5$, *etc.*

$\Delta V_5 = I_5 R_5$

$\approx (12.3\,mA)(500\,\Omega)\ldots$

$\boxed{\Delta V_1 \approx 6.15\,Volts.}$

From Energy Conservation,

It SHOULD be the case that:

$\Delta V_3 \approx \Delta V_2 (\approx \Delta V_2) \approx 2.85\,Volts\ldots$

*Let's Check:*

$\Delta V_2 \approx (4.72\,mA)(600\Omega)$

*ii.* $\boxed{\Delta V_2 \approx 2.83\,Volts \checkmark!}$

*Check:*

$\Delta V_3 \approx (4.04\,mA)(700\ \Omega)$

*iii.* $\boxed{\Delta V_3 \approx 2.83 Volts\ \checkmark!}$

*And, again, this time from the definition of 'parallel configuration',*

*It SHOULD be the case that*

$\Delta V_4 \approx \Delta V_3 \approx \Delta V_2 \approx 1.6mA\ldots$

*Check:*

$\Delta V_4 \approx (3.54\,mA)(800\,\Omega)$

*iv.* $\boxed{\Delta V_4 \approx 2.83\,Volts \checkmark!}$

# IV. **B-Fields (20 pts).**

Be brief but complete and precise regarding all of the following:

A. According to Ampere's historic experimental finding, two light, straight, long wires carrying currents going in parallel directions will do WHAT to each other (1 pts)?

B. Ampere's experimental finding is generally explained by a belief that moving charges create WHAT (1 pts)?

C. Draw a magnetic field line diagram for a long straight current:

  a) Head-On: As though the current is coming out of the whiteboard toward your eye (1 pts).

  b) Side-View: As though the current is traveling in a straight line from one side of the whiteboard toward the other (1 pts).

D. Concisely explain the essential differences between the "Dot Product" and the "Cross Product" for the multiplication of two vectors (3 pts).

E. Write down a clear and complete expression (EQUATION!) for the magnetic field as a function of charge, velocity and displacement from the charge (2 pts).

F. Write down clear and complete expression (EQUATION!) for the magnetic field as a function of current, length and displacement from the current (1 pts).

G. Write down a clear expression for the magnetic force as a function of current, length, and magnetic field (1 pts).

H. Concisely explain the essential differences between the behavior of electric field lines from charges and the behavior of magnetic field lines from charges (3 pts).

IV.B-Fields (20 pts)

A. Two nearby long, light, straight wires carrying parallel currents will *accelerate toward* each other.

B. Moving charges create a *magnetic field*.

C. Below:

c) i.

(1) CONCENTRIC (CLOSED) CIRCLES ENCLOSING THE STRAIGHT CURRENT,

(2) RUNNING COUNTER-CLOCKWISE (right-hand rule)

(3) getting LESS AND LESS DENSE AS THEY get FARTHER AND FARTHER AWAY $(B \propto \frac{1}{r^2})$

ii.

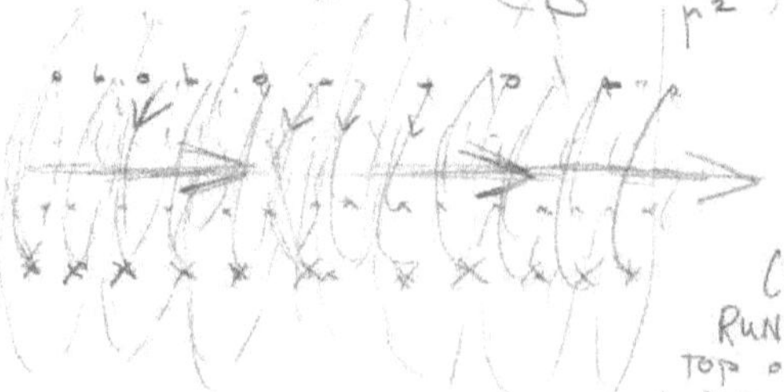

CONCENTRIC CIRCLES RUNNING FROM TOP out of page to BOTTOM in page

D.

The dot product between two vectors (say, vectors $\vec{A}$ and $\vec{B}$) measures the extent to which their directions are aligned; the dot product thereby produces a scalar quantity:

$$\text{Let } C \equiv \vec{A} \cdot \vec{B} \equiv \|\vec{A} \cdot \vec{B}\| \equiv \boxed{AB} \cos\theta$$

Whereas the cross product between two vectors (again, vectors $\vec{A}$ and $\vec{B}$) measures the extent to which their directions differ; the cross product produces a vector quantity:

$$\text{Let } \vec{C} \equiv \vec{A} \times \vec{B}$$
$$\|\vec{C}\| \equiv \boxed{AB} \sin\theta$$
$$\hat{C} \equiv \text{'Right–Hand Rule'}$$

E.

$$d\vec{B} = \frac{\mu_0}{4\pi} \frac{q\vec{v} \times \hat{r}}{r^2}$$

F.

$$d\vec{B} = \frac{\mu_0}{4\pi} \frac{I \cdot \vec{dl} \times \hat{r}}{r^2}$$

G.

$$\vec{F} = I\vec{l} \times \vec{B}$$

H.

The electric field is a vector formed by the multiplication of a 0-Dimensional Scalar source with a vector; the magnetic field is a vector formed from the multiplication of a 1-Dimensional Vector with a vector, a cross-product – for which the resulting direction is perpendicular to both the source current vector and the displacement vector (which is directed from source current to 'field point').

I. The POINT.

a) Draw one long and horizontal current-carrying wire a small amount of space above another identical wire – carrying current in the same direction as the wire below it (1 pt).

b) Referring to your picture, you are going to follow a few specific instructions and, ultimately, provide a thorough physics-based explanation of Oersted's historical observation of two current-carrying wires (and what they apparently do when carefully isolated). . .

REQUIREMENTS/DIRECTIONS

i. The explanation you provide MUST include explicit reference to ***EVERY ONE OF*** your answers provided above (A-H). Make a mark (like an asterisk) at the beginning of the sentence each time you have referred to one of the answers (A-H).

ii. The explanation you provide must be strongly focused on the concept of ***DIRECTION***. You need not be too concerned with the concept of magnitude.

iii. Beginning with a little discussion of the field created by the current above (what does it 'look like'? in what directions do the lines point? etc), proceed step-by-step through discussion of the direction of force exerted by this field and, ultimately, how this force affects the current below.

iv. Supplement your initial drawing with any and every visual detail or side-image that might help clarify your explanation.

The goal of all your descriptions is ultimately to make sense out of Oersted's finding in terms of the laws of physics: specifically, why things 'happen' in the direction that they do. Given your picture and the laws of physics, your explanation must ultimately answer this overall question:

Under highly controlled conditions,

a horizontal current
will be observed to
accelerate up
if a an identically directed current
is located above it

WHY?!

(5 pts).

I.

A)

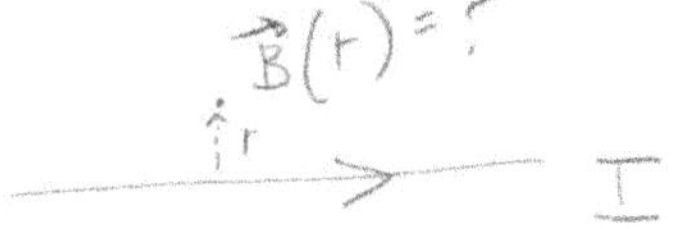


AS CLOSED "AMPERIAN PATH", CHOOSE A CIRCLE of RADIUS $r$ BECAUSE THE MAGNETIC FIELD LINES ARE CIRCLES

SO A CIRCLE of RADIUS $r$ WILL FIND $\vec{B}$-FIELD of CONSTANT MAGNITUDE AND CONSTANT DIRECTION → ALWAYS LYING PERFECTLY PARALLEL TO THE PATH — precisely what the DOT-PRODUCT DEMANDS!

THUS $\oint \vec{B} \cdot d\vec{l} = \mu_0 I_{(enc)}$

$B \oint dl = \mu_0 I_{(enc)}$

B/C THE CHOSEN CLOSED PATH IS A CIRCLE →

$l \equiv$ FULL LENGTH AROUND CLOSED PATH; HERE, CIRCUMFERENCE

CONTINUED

$$B\oint dl = \mu_0 I_{(enc)}$$

$$B \cdot l = \mu_0 I_{(enc)}$$

$$B \cdot 2\pi r = \mu_0 I_{(enc)}$$

$$B \cdot 2\pi r = \mu_0 I$$

So

$$B = \frac{\mu_0}{2\pi} \frac{I}{r}$$

Here, $l$ = CIRCUMF. $= 2\pi r$

$I_{(enc)} \equiv$ ALL CURRENT FLOWING $\perp$ TO AREA BOUND BY CLOSED AMPERIAN PATH

Here, simply $I$

A. (CONTINUED)

$\vec{B}(r) = ?$

FROM BIOT-SAVART (R.H.R. #1)

$$d\vec{B} = \frac{\mu_0}{4\pi} \frac{I\vec{l} \times \hat{r}}{r^2}$$

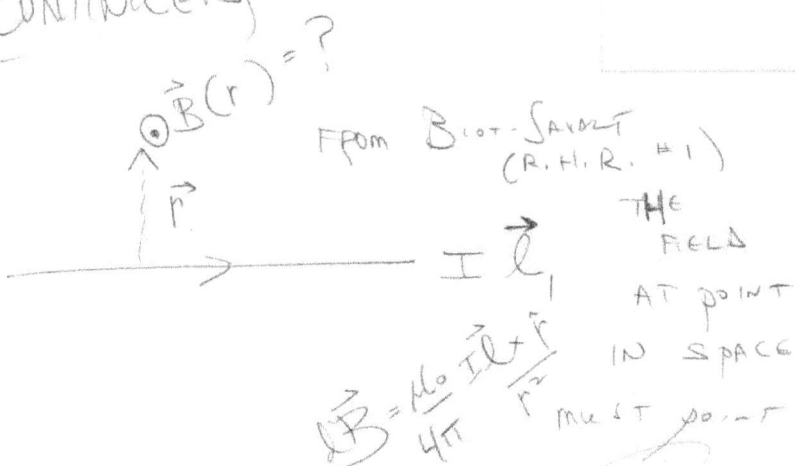

THE FIELD AT POINT IN SPACE MUST POINT OUT w/ R.H. THUMB (FINGERS ALONG $I\vec{l}_1$ CURL UP to $\vec{r}$)

BUT NOW IF WE PUT A NEW CURRENT, $Il_2$ AT THAT OUTWARD POINT FIELD ABOVE $Il_1$ WE APPLY LORENTZ FORCE!

THE FORCE ON $Il_2$ PULLS IT DOWN to $Il_1$!

$\vec{F} = I\vec{l} \times \vec{B}$

FINGERS ALONG $Il_2$ CURL OUT TO FIELD AND THUMB POINTS ↓

(B) SOLENOID

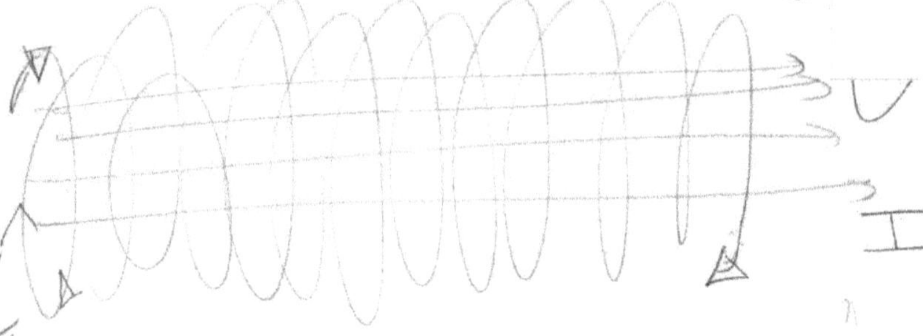

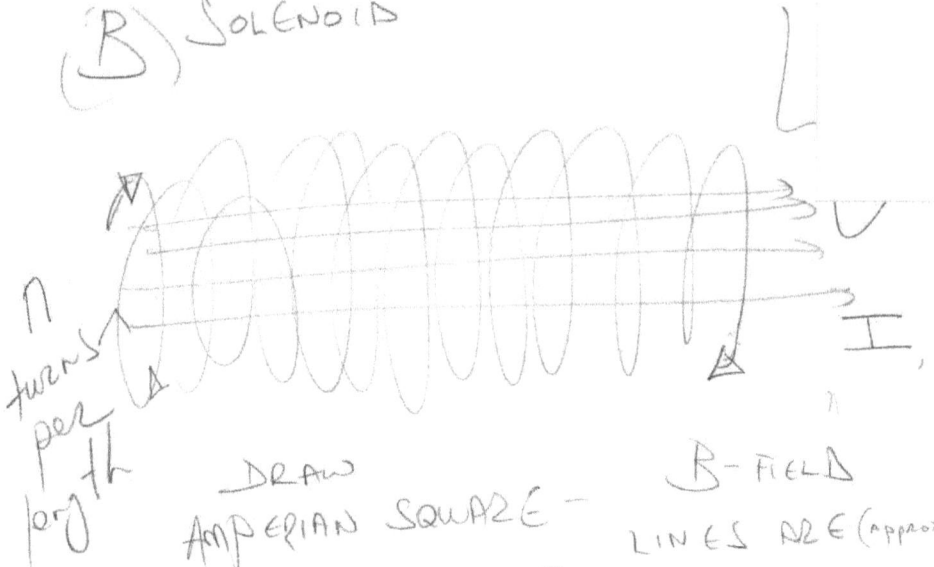

DRAW AMPERIAN SQUARE – HALF IN, HALF OUT of SOLENOID.

Let "S" ≡ length of one side

Let "N" ≡ NUMBER of WIRETURNS IN SQUARE

***

integrate around square counter-clockwise (for CONVENIENCE)

B-FIELD LINES ARE (approx) PARALLEL THROUGH AXIS of SOLENOID; SOLENOID ∞ LONG SO THESE PARALLELL LINES MUST go VERY FAR BEFORE CIRCLING right BACK. ∴ DENSITY of LINES OUTSIDE SOLENOID $\rightarrow 0$

B (CONT'D)
SOLENOID (CONT'D)

CONSIDER THAT Amperian
SQUARE:

THE TWO EDGES THAT ARE
i) ⊥ TO FIELD LINE CONTRIBUTE
0 TO THE INTEGRAL BECAUSE
THE DOT PRODUCT REQUIRES
PARALLEL COMPONENTS (cos θ = 0 here)

ii) the edge OUTSIDE THE SOLENOID
CONTRIBUTES 0 BECAUSE THERE
ARE APPROXIMATELY NO FIELD LINES
THERE

iii) THE ONE EDGE INSIDE SOLENOID
DOES CONTRIBUTE FIELD of
CONSTANT MAGNITUDE AND DIRECTION

So ⟶

B SOLENOID

$$\oint \vec{B} \cdot d\vec{l} = \mu_0 I_{(enc)}$$

$$B \oint dl = \mu_0 I_{(enc)}$$

$$Bl = \mu_0 I_{(enc)}$$

NOW, IN THIS CASE, $l = s$

AND $I_{(enc)} = NI$ (b/c there are a NUMBER of WIRES FLOWING $\perp$ THROUGH THAT SQUARE'S AREA)

SO

$$Bl = \mu_0 N I$$

$$B = \frac{\mu_0 I N}{l}$$

→ we don't know N (depends on our choice of square) and we don't know $l$ (same dependence)

BUT $n \equiv \frac{N}{l}$ = GIVEN SO $\boxed{B = \mu_0 n I}$ — CONSTANT!

no r-dependence vs. b
THE MAGNETIC ne? of a CAP

C

i) put the edge of one solenoid next to the edge of another and you take parallel wires — they'll attract! ↑

ii) TURN ONE SOLENOID AROUND AND THE WIRES ARE ANTIPARALLEL
→ REPULSION!

iii) HUNK OF IRON ≤ SOLENOID
→ MADE OF NATURALLY ALIGNED ORBITALS

iv) "NORTH" = DIRECTION of THUMB!

DIRECTION of B FIELD!

## VI. *Light: The Excluded Middle (20 pts).*

True/False: Put a "T" in the box next to each claim that appears more true than false.

Put an "F" in the box next to each claim that appears more false than true. (1 pt each).

| | |
|---|---|
| 1. As long as a wave's medium does not change in any respect, then the speed of that wave will not change. | T |
| | XXXXXX |
| 2. According to Faraday's Law, if the magnetic flux through some open area does not remain constant in time, then an electric potential will be induced in the closed path bounding that area. | T |
| | XXXXXX |
| 3. ***Nearby but outside*** a charging capacitor ($q > 0$), the magnetic field has a magnitude of 0. | T |
| | XXXXXX |
| 4. ***Inside*** a charging capacitor ($q > 0$), 0 Coulombs per second of electric charge flows from one plate to the other. | F |
| | XXXXXX |
| 5. In a circuit containing a capacitor, no current can flow through the wires until the capacitor is fully charged. | F |
| | XXXXXX |
| 6. Given an open n-dimensional region of space, the region's boundary is a closed region of dimension n-1. | T |
| | XXXXXX |
| 7. Bounded by any closed surface is a unique volume. | T |
| | XXXXXX |

| | |
|---|---|
| 8. Bounded by any closed path is a unique area. | F |
| | XXXXXX |
| 9. If an Amperian loop is drawn outside a charging capacitor, then the area bound by the loop must lie outside the capacitor. | F |
| | XXXXXX |
| 10. Inside a charging capacitor, 0 magnitude of electric flux flows from one plate to the other. | F |

| | |
|---|---|
| | XXXXXX |
| 11. Maxwell's Displacement Current correction was added to Gauss's Law and allowed Gauss's Law to apply to steady currents. | F |
| | |
| 12. According to the explanation of Maxwell's Displacement Current, if the magnitude of an electric flux through an open area changes in time, a magnetic field will be induced in the closed path bounding that area. | T |
| | XXXXXX |
| 13. Mutually induced electric and magnetic fields must be perpendicular to each other. | T |
| | XXXXXX |
| 14. Given the function $I(t) = \frac{1}{6}at^3$, no matter how many times you differentiate with respect to t ($\frac{dI}{dt}, \frac{d^2I}{dt^2}, \frac{d^3I}{dI^3}, \ldots$), you will never arrive at the answer "0". | F |
| | XXXXXX |
| 15. Given the function $I(t) = \frac{1}{6}e^{3t}$, no matter how many times you differentiate with respect to t ($\frac{dI}{dt}, \frac{d^2I}{dt^2}, \frac{d^3I}{dI^3}, \ldots$), you will never arrive at the answer "0". | T |
| | XXXXXX |
| 16. If you find charge in some lab, you will necessarily find electric and/or magnetic field in that lab. | T |
| | XXXXXX |
| 17. If you find electric and/or magnetic field in some lab, you will necessarily find charge in that lab. | F |
| | XXXXXX |
| 18. When electric and magnetic fields mutually induce each other in a perpetual pattern, that pattern satisfies the wave equation. | T |
| | XXXXXX |
| 19. $\frac{1}{\sqrt{\varepsilon_0\mu_0}}$ = the speed at which mutually inducing electric and magnetic fields propagate through a vacuum. | T |
| | XXXXXX |
| 20. $\frac{1}{\sqrt{\varepsilon_0\mu_0}}$ = the speed at which light travels through a vacuum. | T |
| | XXXXXX |
| 21. *** BONUS *** You will live long and prosper; the net force will be with you. | T |

# PART II

SPRING 2017

3

# FULL-TEXT OF FINAL EXAM

# S17: BLANK

# Physics 204, S17
# Final Exam

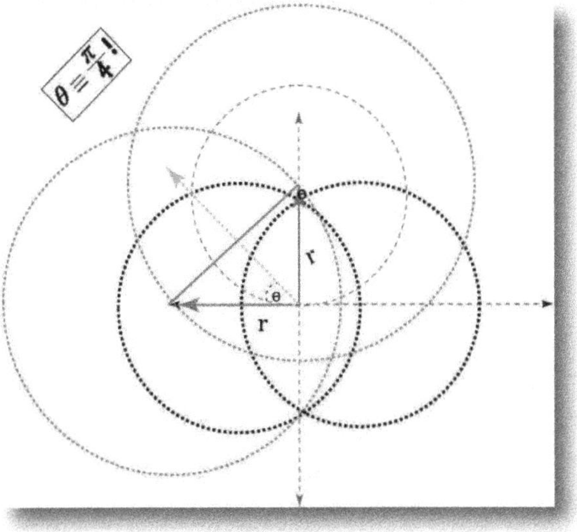


# re Flux
# and related matters.

John Jay College of Criminal Justice
The City University of New York
Walters, Wu, Martens Yaverbaum
Week of Monday, May 22, 2017

## SOME USEFUL RELATIONS:

1) $\oint \vec{E} \cdot \overrightarrow{da} = \frac{q_{(enc)}}{\varepsilon_0}$.

2) $\oint \vec{B} \cdot \overrightarrow{da} = 0$.

3) $\oint \vec{E} \cdot \overrightarrow{dl} = -\frac{d}{dt} \int \vec{B} \cdot \overrightarrow{da}$.

4) $\oint \vec{B} \cdot \overrightarrow{dl} = \mu_0 I_{(enc)} + \mu_0 \varepsilon_0 \frac{d}{dt} \int \vec{E} \cdot \overrightarrow{da}$.

5) $\vec{E} = \frac{1}{4\pi\varepsilon_0} \frac{q}{r^2} \hat{r}$.

6) $\vec{F}_E \equiv q\vec{E}$.

7) $\vec{B} = \frac{\mu_0}{4\pi} \frac{q}{r^2} \vec{v} \times \hat{r}$.

8) $\vec{F}_B \equiv q\vec{v} \times \vec{B} = \vec{I}l \times \vec{B}$.

9) $V_a - V_b \equiv \int_a^b \vec{E} \cdot \overrightarrow{dr}$.

10) $C \equiv \frac{Q}{\Delta V}$. (Note: This "C" is capital.)

11) $I \equiv \frac{dq}{dt}$

12) $I = \frac{\Delta V}{R}$.

13) $\mathcal{E} - IR - \frac{Q}{C} = 0$. (Capital "C".)

14) $c \approx 3 \; x \; 10^8 \; m/s$.

15) $n \equiv \frac{c}{v}$. (Lower-case "c".)

16) $n_1 \sin\theta_1 = n_2 \sin\theta_2$.

17) $\frac{Yd}{L} = \frac{n\lambda}{2}$

18) $sin^2\theta + cos^2\theta = 1.$

19) $\bar{v} \equiv \frac{\Delta x}{\Delta t}.$

20) $\sum \vec{F} = m\vec{a}.$

21) $F = -Kx.$

22) $x = Acos(\omega t + \phi).$

23) $\omega = 2\pi f.$

24) $f = \frac{1}{T}.$

25) $KE = \frac{1}{2}mv^2$

26) $PE_{elastic} = \frac{1}{2}Kx^2$

27) $v = \lambda f.$

28) $v = \sqrt{\frac{T}{\mu}}.$

29) $\frac{\partial^2 y}{\partial t^2} = (v^2)\frac{\partial^2 y}{\partial x^2}.$

30) $\epsilon_0 \approx 8.85 \, x \, 10^{-12} \, \frac{C^2}{Nm^2}.$

31) $\mu_0 \approx 1.26 \times 10^{-7} \, \frac{N}{A^2}.$

32) $m_e \approx 9.11 \, x \, 10^{-31}$ kg.

33) $m_{(NEUTRON)} \approx 1.67 \, x \, 10^{-27}$ kg

34) $q_e \approx 1.60 \, x \, 10^{-19} \, C.$

Two ***point-charges*** of differing magnitudes are held stationary in an enormously spacious x-y plane.

A researcher places an instrument called a 'field detector' at the point **(-1,-1)**.

The detector is designed to measure electrostatic field magnitudes and directions. This one, you may assume, did so with stunningly high precision.

Prior to all this, some standard measurements were made, and characteristic data regarding the two point charges were organized into a table, shown below.

| Name (formal, alias) | Charge (Coulombs) | x-Coordinate (meters) | y-Coordinate (meters) | Ordered Pair (x, y) |
|---|---|---|---|---|
| ('Qi Wu') **$Q_1$** | **$+5 \times 10^{-10}$** | **-7** | **9** | **(-7, 9)** |
| ('Jo Wa') **$Q_2$** | **$-12 \times 10^{-10}$** | **17** | **23** | **(17, 23)** |

Note: All coordinates are measured and given in **meters** (not centimeters); similarly, the (enormous) charge magnitudes are in whole **Coulombs** (not micro-Coulombs).

Also Note: If you wish, you are permitted and encouraged to approximate the electrostatic constant as:

$$K_e \approx 10 \times 10^9 \approx 1 \times 10^{10} \frac{Nm^2}{C^2}.$$

* * *

a) Draw a neat and clear sketch of the situation, as you understand it. Your sketch must express a clear decision as to which directions are designated by + and – on each axis (4 pts).

b) For *each* of the two individual point charges, draw an approximate field line diagram – each drawn as it would look IF the other charge did NOT exist. For BOTH, however, obey the following convention: For every $1 \times 10^{-10}$ Coulombs of strength that generates a field, at least one more field line should appear in the field line diagram (2 pts each).

c) Compute the Electric Field as measured at the *Point of Interest* : (-1,-1). That is:

i. In Newtons/Coulomb, determine the ***magnitude*** of the ('total') ***electrostatic field*** at this location of interest (7 pts).

ii. Using degrees **and** points of the compass (such as "20 degrees South-West"), express the **direction End Sub** of the ***electrostatic field*** at this *Point of Interest* (-1,-1) (3 pts).

Now, suppose that at some given instant, we notice
ONE ELECTRON held stationary
at the *Point of Interest*: (-1,-1).
At a moment we will call ***t = 0***,
the electron is released from rest at that location.

iii. Find (a) the magnitude (3 pts) and
(b) the direction (2 pts) of
the electron's immediate instantaneous acceleration,
upon finding itself in the electrostatic field described above.

iv. TRUE or FALSE (?): |

a) In general, an electron in a field accelerates in a direction ***opposite*** to that of the field lines (1 pt).

b) in general, an electron in a field tends to select
and travel along a single field line
until it encounters a proton, infinity or the ground (1 pt).

A huge but imaginary sphere is drawn with its center at the origin of the coordinate system used in Problem I, above.
The radius of the sphere is $\boldsymbol{r}$ = ***30 meters***.

**ALL INFORMATION given or found in the above Problem I STILL APPLIES- but the electron (test charge) is removed.**

The huge and imaginary sphere contains both of the charges given in Problem I. And nothing else.

a) *Draw this situation* as you understand it (4 pts).

(This is not meant to be a trick; the drawing will just help clarify.)

b) In $\frac{Newtons \cdot Meters^2}{Coulombs}$, find the total amount of ***electric flux*** that passes through the surface of this imaginary sphere.

Your answer should be a number, expressed in the units mentioned directly above (6 pts).

c) The two point charges from Problem I are replaced with an extremely long and straight ***wire*** (i.e.: LINE) that is net-positively charged; this straight line wire stretches through the points (-7, 9) and (17, 23) and beyond.

The charge density in the wire is constant and expressed as follows, where $\boldsymbol{L}$ refers to length measured in meters and $\boldsymbol{Q}$ is charge measured in Coulombs:

$$\lambda \equiv \frac{+Q}{L}$$

i. Draw the field line diagram produced by this long line of positive charge (6 pts).

ii. In a sentence or two of English, explain why the sphere drawn above will not be the most convenient shape for computing ***electric field*** at $\boldsymbol{r}$ (3 pts).

iii. Choose and draw a Gaussian (closed) surface surrounding some portion of the line—a surface that will best help you compute the field at $\boldsymbol{r}$ (6 pts).

iv. Starting with Gauss's Law and proceeding through as many clear and verbally explained steps as possible,

find $\boldsymbol{E}$ as a function of $\boldsymbol{r}$,

where $\boldsymbol{E}$ refers to the magnitude and direction of the electrostatic field at some point in space, and where the point in space is understood to be influenced by the nearby presence of a long, thin ***line*** of charge: The line, we may assume, is of constant (1-D) charge density $\boldsymbol{\lambda}$ and is located a displacement of $\boldsymbol{r}$ away from the point in space. Your answer will not consist of numbers. It will consist largely of letters, and, ultimately, of no more than the letters mentioned above (10 pts).

## III. An actual CIRCUIT (25 pts).

Examine the following circuit (values provided directly to its right).

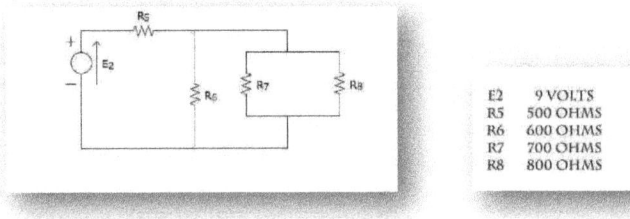

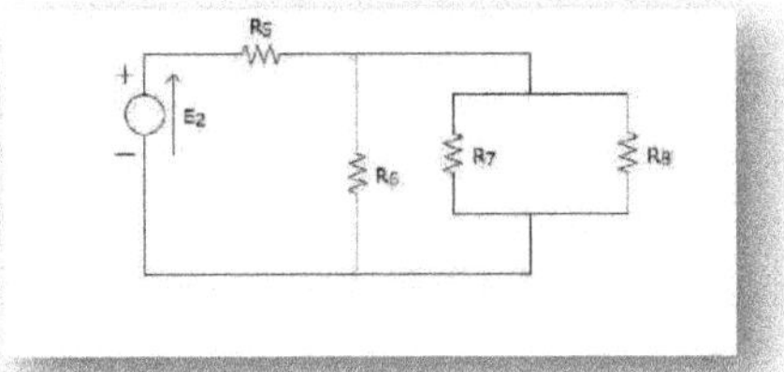


| | |
|---|---|
| E2 | 9 VOLTS |
| R5 | 500 OHMS |
| R6 | 600 OHMS |
| R7 | 700 OHMS |
| R8 | 800 OHMS |

SHOWING ALL WORK, Determine:

a) The ***current*** flowing through each and every RESISTOR (3 pts each).

i. ***I*** (R5) =

ii. ***I*** (R6) =

iii. ***I*** (R7) =

iv. ***I*** (R8) =

b) The ***potential difference*** across each and every RESISTOR (3 pts each, +1 pt for being you).

v. ***Δ V*** (R5) =

vi. ***Δ V*** (R6) =

vii. ***Δ V*** (R7) =

viii. ***Δ V*** (R8) =

IVV. THE B-FIELD. DIRECTION (15 pts).

**Let:**

$$\vec{A} \equiv 3\hat{x} - 5\hat{y} + \hat{z}$$
$$\vec{B} \equiv 6\hat{x} + 5\hat{y} + \hat{z}$$
$$\vec{C} \equiv 4$$

A. Find $\left(\vec{\mathbf{B}} \cdot \vec{\mathbf{A}}\right)$ (2 pts.)

B. Find $\vec{\mathbf{B}} \times \left(\vec{\mathbf{B}} \times \vec{\mathbf{A}}\right)$ (2 pts.)

C. Find $c\left(\vec{\mathbf{A}} \times \vec{\mathbf{B}}\right)$ (2 pts.)

Let the magnetic field produced by an infinitesimal amount of current
be given by the Biot-Savart Law:

$$d\vec{\mathbf{B}} = \frac{\mu_0}{4\pi}\frac{I\vec{dl}}{r^2} \times \hat{r}$$

and

Let the magnetic force exerted on a correspondingly small amount of current
by the magnetic field (produced in the manner described above) be

$$\vec{\mathbf{F}}_B = I\vec{dl} \times \vec{\mathbf{B}}$$

and let:

The direction of the vector produced by the 'cross' multiplication of two vectors,
If not otherwise adjudicated, be indicated by

***The 'Right-Hand' Rule.***

**Then:**

**D) Underneath a long straight current flowing to the right, in what direction does the magnetic field point (3 pts)?**

**E) If a rightward-flowing current is placed into a magnetic field which itself is pointing into the board, in what direction will the current be*forced (2 pts)* ?**

**F) A metal rod faces North/South on two metal rails that run East/West. There is a resistor running North/South and clamped down to complete a metal rectangle. Some sort of mechanical force is applied to the metal rod so that it is pulled Eastward along the rails and thereby progressively expands the area of the rectangle. The whole apparatus is submerged in a constant magnetic field which points up toward the ceiling.**

**1) In what direction will positive charges in the rod be forced to move (North? West? (2 pts))**

**2) In what direction will current necessarily begin to flow? (clockwise?counterclockwise? (2 pt))**

# 4
# FULL-TEXT OF FINAL EXAM S17: SOLVED AND EXPLAINED

SOLVED

# Physics 204, S17
## Final Exam

$\theta \equiv \frac{\pi}{4}$!

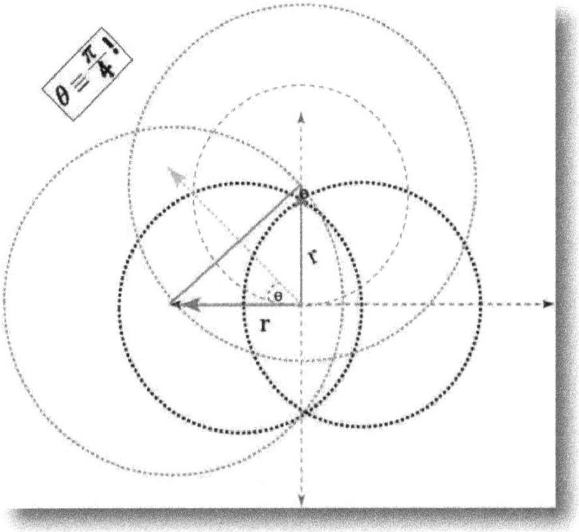


# re Flux and related matters.

John Jay College of Criminal Justice
The City University of New York
Walters, Wu, Martens Yaverbaum
Week of Monday, May 22, 2017

## SOME USEFUL RELATIONS:

1) $\oint \vec{E} \cdot \overrightarrow{da} = \frac{q_{(enc)}}{\varepsilon_0}$.

2) $\oint \vec{B} \cdot \overrightarrow{da} = 0$.

3) $\oint \vec{E} \cdot \overrightarrow{dl} = -\frac{d}{dt} \int \vec{B} \cdot \overrightarrow{da}$.

4) $\oint \vec{B} \cdot \overrightarrow{dl} = \mu_0 I_{(enc)} + \mu_0 \varepsilon_0 \frac{d}{dt} \int \vec{E} \cdot \overrightarrow{da}$.

5) $\vec{E} = \frac{1}{4\pi\varepsilon_0} \frac{q}{r^2} \hat{r}$.

6) $\vec{F}_E \equiv q\vec{E}$.

7) $\vec{B} = \frac{\mu_0}{4\pi} \frac{q}{r^2} \vec{v} \times \hat{r}$.

8) $\vec{F}_B \equiv q\vec{v} \times \vec{B} = \vec{I}l \times \vec{B}$.

9) $V_a - V_b \equiv \int_a^b \vec{E} \cdot \overrightarrow{dr}$.

10) $C \equiv \frac{Q}{\Delta V}$. (Note: This "C" is capital.)

11) $I \equiv \frac{dq}{dt}$

12) $I = \frac{\Delta V}{R}$.

13) $\mathcal{E} - IR - \frac{Q}{C} = 0$. (Capital "C".)

14) $c \approx 3 \, x \, 10^8 \, m/s$.

15) $n \equiv \frac{c}{v}$. (Lower-case "c".)

16) $n_1 \sin\theta_1 = n_2 \sin\theta_2$.

17) $\frac{Yd}{L} = \frac{n\lambda}{2}$

18) $sin^2\theta + cos^2\theta = 1.$

19) $\bar{v} \equiv \frac{\Delta x}{\Delta t}.$

20) $\sum \vec{F} = m\vec{a}.$

21) $F = -Kx.$

22) $x = Acos(\omega t + \phi).$

23) $\omega = 2\pi f.$

24) $f = \frac{1}{T}.$

25) $KE = \frac{1}{2}mv^2$

26) $PE_{elastic} = \frac{1}{2}Kx^2$

27) $v = \lambda f.$

28) $v = \sqrt{\frac{T}{\mu}}.$

29) $\frac{\partial^2 y}{\partial t^2} = (v^2)\frac{\partial^2 y}{\partial x^2}.$

30) $\epsilon_0 \approx 8.85 \, x \, 10^{-12} \, \frac{C^2}{Nm^2}.$

31) $\mu_0 \approx 1.26 \times 10^{-7} \, \frac{N}{A^2}.$

32) $m_e \approx 9.11 \, x \, 10^{-31}$ kg.

33) $m_{(NEUTRON)} \approx 1.67 \, x \, 10^{-27}$ kg

34) $q_e \approx 1.60 \, x \, 10^{-19} \, C.$

Two ***point-charges*** of differing magnitudes are held stationary in an enormously spacious x-y plane.

A researcher places an instrument called a 'field detector' at the point **(-1,-1)**.

The detector is designed to measure electrostatic field magnitudes and directions. This one, you may assume, did so with stunningly high precision.

Prior to all this, some standard measurements were made, and characteristic data regarding the two point charges were organized into a table, shown below.

| Name (formal, alias) | Charge (Coulombs) | x-Coordinate (meters) | y-Coordinate (meters) | Ordered Pair (x, y) |
|---|---|---|---|---|
| ('Qi Wu') $Q_1$ | $+5 \times 10^{-10}$ | -7 | 9 | (-7, 9) |
| ('Jo Wa') $Q_2$ | $-12 \times 10^{-10}$ | 17 | 23 | (17, 23) |

Note: All coordinates are measured and given in **meters** (not centimeters); similarly, the (enormous) charge magnitudes are in whole **Coulombs** (not micro-Coulombs).

Also Note: If you wish, you are permitted and encouraged to approximate the electrostatic constant as:

$$K_e \approx 10 \times 10^9 \approx 1 \times 10^{10} \frac{Nm^2}{C^2}.$$

* * *

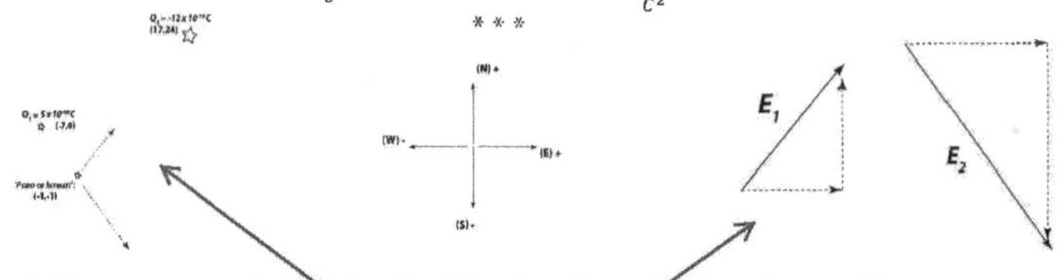


a) Draw a neat and clear sketch of the situation, as you understand it. Your sketch must express a clear decision as to which directions are designated by + and – on each axis (5 pts).

b) For *each* of the two individual point charges, draw an approximate field line diagram – each drawn as it would look IF the other charge did NOT exist. For BOTH, however, obey the following convention: For every $1 \times 10^{-10}$ Coulombs of strength that generates a field, at least one more field line should appear in the field line diagram (5 pts).

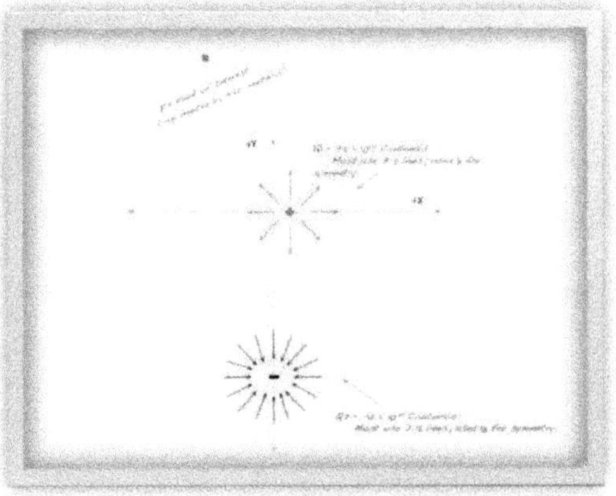
+Y
+X

c) Compute the Electric Field as measured at the *Point of Interest* : (-1,-1).

That is:

i. In Newtons/Coulomb, determine the ***magnitude*** of the ('total') ***electrostatic field*** at this location of interest (3 pts).

ii. Using degrees **and** points of the compass (such as "20 degrees South-West"), express the **direction End Sub** of the ***electrostatic field*** at this *Point of Interest* (-1,-1) (2 pts).

Now, suppose that at some given instant, we notice
ONE ELECTRON held stationary
at the *Point of Interest*: (-1,-1).
At a moment we will call $t = 0$,
the electron is released from rest at that location.

iii. Find (a) the magnitude (3 pts) and (b) the direction (2 pts)

of the electron's immediate instantaneous acceleration, upon finding itself in the electrostatic field described above.

iv. TRUE or FALSE (?): |

a) In general, an electron in a field accelerates in a direction ***opposite*** to that of the field lines (3 pts)

b) In general, an electron in a field tends to select and travel along a single field line until it encounters a proton, infinity or the ground (2 pts).

**P204.S17.Final Exam**
**Problem 1(c). updated 5 − 27, 11'08pm**

$$E_{1x} = \frac{kQ_1}{r^2}\cos\theta$$
$$= \frac{\pm(1\times10^{10})(5\times10^{-10})}{36+100}\left(\frac{6}{\sqrt{(36+100)}}\right)\varepsilon$$
$$= \frac{\pm 30}{(136)^{3/2}}$$

$Q_1$ is positive and to the left of the point of interest, so its field contribution points to the right +

$$\approx \frac{+30}{1586}$$
$$E_{1x} \approx +.0189$$

$$E_{1y} = \frac{kQ_1}{r^2}\sin\theta$$
$$= \pm\frac{5}{136}\left(\frac{10}{\sqrt{136}}\right)$$
$$E_{1y} = \pm\frac{50}{(136)^{3/2}}$$

$Q_1$ is positive and above point of interest, so its field contribution points down −

$$\approx \frac{-50}{1586}$$
$$\approx \frac{-50}{1586}$$
$$\approx -.0315$$

$$E_{2x} = \frac{kQ_2}{r^2}\cos\theta$$
$$= \pm\frac{(12)}{324+576}\frac{18}{\sqrt{324+576}}$$
$$= \pm\frac{216}{(900)^{3/2}}$$
$$= \pm\frac{216}{27000}$$

$Q_2$ [illegible]

$$E_{2x} \approx +.00800$$
$$E_{2y} = \frac{kQ_2}{r^2}\sin\theta$$
$$= \pm\frac{(12)}{324+576}\frac{24}{\sqrt{324+576}}$$

$Q_2$ [illegible]

$$= +\frac{288}{(900)^{3/2}}$$
$$= +\frac{288}{27000}$$
$$E_{2y} \approx +.0107$$

So $\sum E_x \approx +.0189 + (+.008)$
$\sum E_x \approx +.0269$

$\sum E_y \approx -.0315 + (+.0107)$
$\sum E_y \approx -.0208$

$$E \approx \sqrt{(.0269)^2 + (.0208)^2}$$

$$\approx .0340$$

$$\theta \approx \tan^{-1}\left(\frac{.0208}{.0269}\right)$$

$$\approx$$

c)

i.

$$E \approx .0340 \text{ N/C}$$

ii.

$$\theta \approx 37.7^\circ \text{ S} - \text{E}$$

iii.

$$\vec{F} = q\vec{E}$$

$$\sum \vec{F} = m\vec{a}$$

$$F = (-1.60 \times 10^{-19} \text{C})(.0340 \text{ N/C})$$

$$F = 5.44 \times 10^{-21} \text{N}$$

$$a = \frac{5.44 \times 10^{-21} \text{N}}{9.11 \times 10^{-31} \text{kg}}$$

(a) $a \approx 5.97 \times 10^{9} \text{m/s}^2$

(b) $\theta \approx 37.7^\circ \text{ N} - \text{W}$

iv.

(a) TRUE

(b) FALSE

| | |
|---|---|
| $E_{1x} \approx +1.89 \times 10^{-2} N/C$ | $E_{1y} \approx -3.15 \times 10^{-2} N/C$ |
| $E_{2x} \approx +8.00 \times 10^{-3} N/C$ | $E_{2y} \approx +1.07 \times 10^{-2} N/C$ |
| | |
| $E_x \approx +2.69 \times 10^{-2} N/C$ | $E_y \approx -2.08 \times 10^{-2} N/C$ |
| | |
| $E \approx 3.40 \times 10^{-2} N/C$ | $\theta \approx 37.7° S-E$ |
| | |
| $F \approx 5.44 \times 10^{-21} N$ | $a \approx 5.97 \times 10^{9} m/s^2$ |

A huge but imaginary sphere is drawn with its center at the origin of the coordinate system used in Problem I, above.
The radius of the sphere is $r$ = ***30 meters***.

**ALL INFORMATION given or found in the above Problem I STILL APPLIES- but the electron (test charge) is removed.**

The huge and imaginary sphere contains both of the charges given in Problem I.
And nothing else.

a) *Draw this situation* as you understand it (4 pts).

(This is not meant to be a trick; the drawing will just help clarify.)

b) In $\frac{Newtons \cdot Meters^2}{Coulombs}$, find the total amount of ***electric flux*** that passes through the surface of this imaginary sphere.

Your answer should be a number, expressed in the units mentioned directly above (6 pts).

c) The two point charges from Problem I are replaced with an extremely long and straight ***wire*** (i.e.: LINE) that is net-positively charged; this straight line wire stretches through the points (-7, 9) and (17, 23) and beyond.

The charge density in the wire is constant and expressed as follows, where $L$ refers to length measured in meters and $Q$ is charge measured in Coulombs:

$$\lambda \equiv \frac{+Q}{L}$$

i. Draw the field line diagram produced by this long line of positive charge (6 pts).

ii. In a sentence or two of English, explain why the sphere drawn above will not be the most convenient shape for computing ***electric field*** at $r$ (3 pts).

iii. Choose and draw a Gaussian (closed) surface surrounding some portion of the line—a surface that will best help you compute the field at $r$ (6 pts).

iv. Starting with Gauss's Law and proceeding through as many clear and verbally explained steps as possible,

find $E$ as a function of $r$,

where $E$ refers to the magnitude and direction of the electrostatic field at some point in space, and where the point in space is understood to be influenced by the nearby presence of a long, thin ***line*** of charge: The line, we may assume, is of constant (1-D) charge density $\lambda$ and is located a displacement of $r$ away from the point in space. Your answer will not consist of numbers. It will consist largely of letters, and, ultimately, of no more than the letters mentioned above (10 pts).

a)

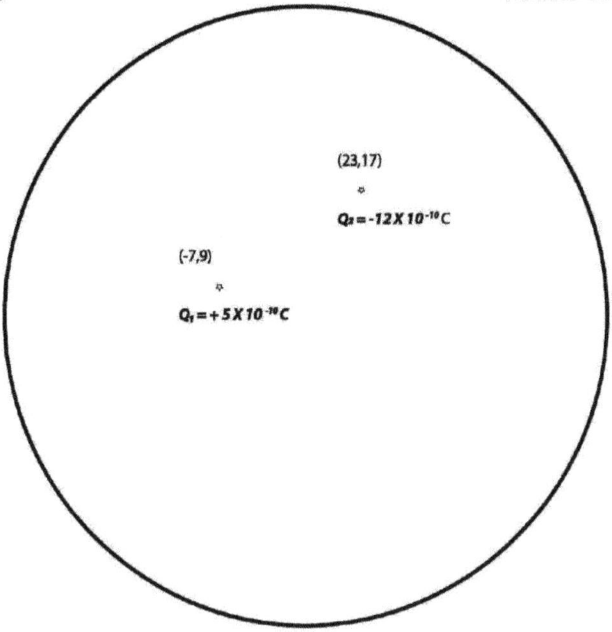

b) In $\frac{Newtons \cdot Meters^2}{Coulombs}$, find the total amount of ***electric flux*** that passes through the surface of this imaginary sphere (2 pts).

Your answer should be a number, expressed in the units mentioned directly above.

*Definition of Electrostatic FLUX:*

$$\Phi_E \equiv \oint \vec{\mathbf{E}} \cdot d\vec{\mathbf{A}}.$$

*Gauss's Law regarding Electrostatic FLUX:*

$$\oint \vec{\mathbf{E}} \cdot d\vec{\mathbf{A}} = \frac{q_{(enc)}}{\varepsilon_0}$$

(The electrostatic flux through ANY closed surface
is always and simply the total magnitude of the charge
enclosed by that surface – –divided by a constant
(known as the 'permittivity of free space') – –
entirely independent of the size or shape
of the closed surface,
entirely independent of the shape
of the enclosed charge
and
entirely independent of ANY charges located anywhere
OUTSIDE the closed surface.
(!)
So, the answer is simply:

$$\frac{(5.00 \times 10^{-10}\,Coulombs) - (12.0 \times 10^{-10}\,Coulombs)}{\varepsilon_0}$$

$$\approx \frac{-7.00 \times 10^{-10}\,Coulombs}{8.85 \times 10^{-12}\,C^2/_{N \cdot m^2}}$$

$$\boxed{\Phi_E = q_{(enc)} \approx -79.1\,Nm^2/C!}$$

c) The two point charges from Problem I are replaced with an extremely long and straight ***wire*** (i.e.: LINE) that is net-positively charged; this straight line wire stretches through the points (-7,9) and (17,23,) and beyond.

The charge density in the wire is constant and expressed as follows, where ***L*** refers to length measured in meters and ***Q*** is charge measured in Coulombs:

$$\lambda \equiv \frac{\pm Q}{L}.$$

i. Draw the field line diagram produced by this long line of positive charge (6 pts).

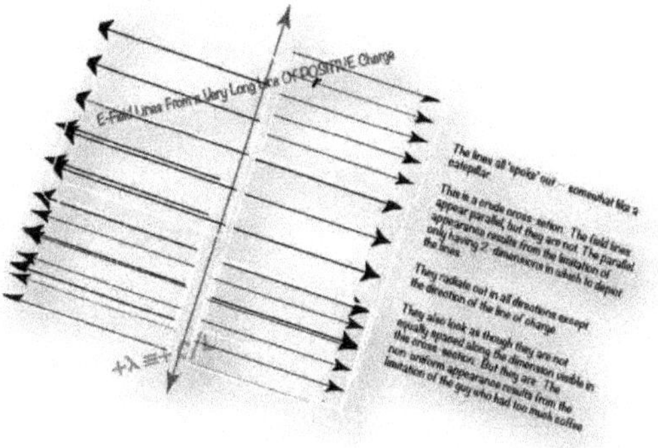


ii. In a sentence or two of English, explain why the sphere drawn above will not be the most convenient shape for computing ***electric field*** at $\boldsymbol{r}$ (6 pts).

Gauss's Law is true for ANY closed surface, so a sphere CAN absolutely be drawn around a vertically oriented continuous line of charge. The total flux through the surface of that sphere WILL be the total charge $(-\lambda L)$ enclosed by that sphere - despite the fact that a sphere produces field lines that are symmetric in all directions with NO exception ("0-dimensional symmetry") while a line is symmetric in all directions save ONE (its own axis - hence "1-dimensional symmetry"). Truth is not the problem. The problem is that this difference in pattern (between a line and a sphere) means that the lines spoking from the line of charge will all 'flow' (***flux***) through the sphere surface at different angles and in densities that will differ from region to region. The magnitude and the direction of the field lines, that is, will NOT BE A CONSTANT through the surface of that sphere. So the E-Field term in the Flux Integral CANNOT be 'taken out' of the integral and we actually have to compute an integral. And not an easy one, to say the least. A central purpose of Gauss's Law is to avoid cumbersome or intractable integral calculations by thinking conceptually.... Physically instead. To trade math for physics. If we draw a sphere around the line of charge, Gauss's Law is still right, but it just isn't helpful. We want to set up an integral that we do NOT actually have to do.

iii. Choose and draw a Gaussian (closed) surface surrounding some portion of the line—a surface that will best help you compute the field at $\boldsymbol{r}$ (6 pts).|

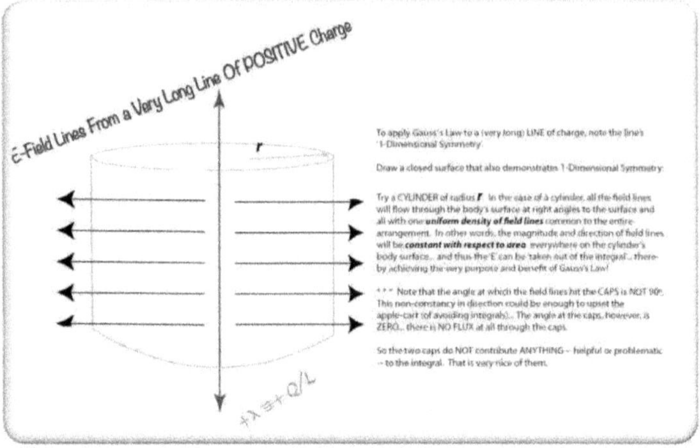
E-Field Lines From a Very Long Line Of POSITIVE Charge
r
+λ = +Q/L
To apply Gauss's Law to a (very long) LINE of charge, note the line's '1-Dimensional Symmetry'.
Draw a closed surface that also demonstrates 1-Dimensional Symmetry:
Try a CYLINDER of radius r. In the case of a cylinder, all the field lines will flow through the body's surface at right angles to the surface and all with one uniform density of field lines common to the entire arrangement. In other words, the magnitude and direction of field lines will be constant with respect to area everywhere on the cylinder's body surface... and thus the 'E' can be 'taken out of the integral'... thereby achieving the very purpose and benefit of Gauss's Law!
* * * Note that the angle at which the field lines hit the CAPS is NOT 90°. This non-constancy in direction could be enough to upset the apple-cart (of avoiding integrals)... The angle at the caps, however, is ZERO... there is NO FLUX at all through the caps.
So the two caps do NOT contribute ANYTHING -- helpful or problematic -- to the integral. That is very nice of them.

iv. Starting with Gauss's Law and proceeding through as many clear and verbally explained steps as possible,

find $\boldsymbol{E}$ as a function of (perpendicular distance) $\boldsymbol{r}$
(given $\boldsymbol{\lambda}$, $\boldsymbol{\varepsilon_0}$)
from this line of charge 7 pts).

This answer will not be a number.

$$\oint \vec{\mathbf{E}} \cdot d\vec{\mathbf{A}} = \frac{q_{(enc)}}{\varepsilon_0}$$

$$E \oint dA = \frac{q_{(enc)}}{\varepsilon_0}$$

(See explanation in 2 (b) ii, above.)

NOTE: On the one hand, if you cannot legitmately make this simplification,

then Gauss's Law does you no good.

Applied properly, the E will ALWAYS 'come out' of the integral.

On the other hand, you cannot simply skip over all the drawing

and over the CHOOSING of the most convenient closed ('Gaussian') surface to draw:

It might well seem as though all the drawing and choosing is all nonsense - -

that we should just write down $EA = \frac{q(enc)}{\varepsilon_0}$ and get on with it

since it's apparently going to be true every time.

*... But Gauss was no fool. Neither be you! Should you wish to drain the bathwaters of thick integral computation, then clutch tightly the babies of well-chosen surface and meaningful integral set-up! ... →*

On the contrary: The drawing and choosing is the heart of the method.

Without drawing and choosing, you don't know WHAT SURFACE AREA is being used

(supposedly by you!) and thus you DO NOT KNOW by what expression you will divide FLUX

in order to get FIELD - -

Determination of the FIELD WAS AND IS THE GOAL of considering Flux. So... back to this calculation

of the E - Field at a distance r from a very long line of charge, . . .

$$\ldots E \oint dA = \frac{q_{(enc)}}{\varepsilon_0}$$

$$EA = \frac{q_{(enc)}}{\varepsilon_0}$$

In this particular example/application, therefore,

'area' refers to the surface area of a CYLINDER (radius r, length L):

Surface Area(cylinder) = $2\pi rL$.

$$E \cdot 2\pi rL = \frac{q_{(enc)}}{\varepsilon_0}$$

$$E = \frac{1}{2\pi\varepsilon_0}\frac{q}{L}\frac{1}{r} \ldots \textit{by definition, then,} \ldots$$

$$\boxed{E = \frac{1}{2\pi\varepsilon_0}\frac{\lambda}{r}!}$$

## III. An actual CIRCUIT (25 pts).

Examine the following circuit (values provided directly to its right).

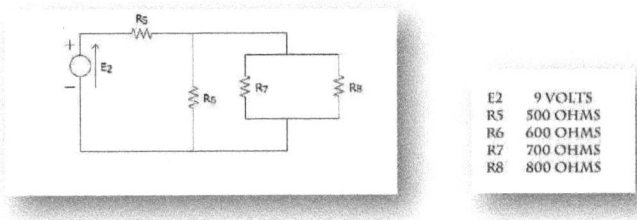

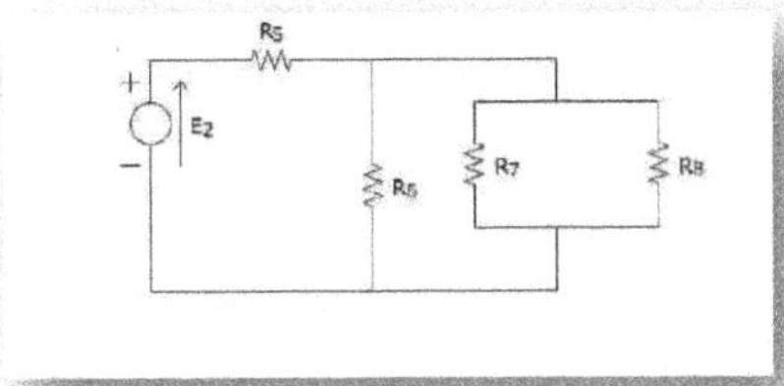


| | |
|---|---|
| E2 | 9 VOLTS |
| R5 | 500 OHMS |
| R6 | 600 OHMS |
| R7 | 700 OHMS |
| R8 | 800 OHMS |

SHOWING ALL WORK, Determine:

a) The ***current*** flowing through each and every RESISTOR (3 pts each.

i. ***I*** (R5) =

$$\frac{1}{R_{eq(R6,R7,R8)}} = \frac{1}{600} + \frac{1}{700} + \frac{1}{800}$$

$$\frac{1}{R_{eq(R2,R3,R4)}} \approx 4.34 \times 10^{-3}\ Ohms^{-1}$$

$$R_{eq(R2,R3,R4)} \approx 230\ Ohms$$

*Therefore,*

$$R_{eq(R1,R2,R3,R4)} \approx 500\ Ohms + 230\ Ohms$$

$$R_{eq(R1,R2,R3,R4)} \approx 730\ \Omega$$

$$I_{Battery} \equiv I_{MainLoop} \equiv \frac{\varepsilon}{R_{eq}}$$

*Here,* $I_5 = I_{Battery}$

$$I_5 \approx \frac{9\ Volts}{730\ Ohms} \approx 1.23 \times 10^{-2}\ Amperes$$

$$\boxed{I_5 \approx 12.3\ milliAmperes}$$

ii. $\boldsymbol{I}$(R6) = . . .

Now, between ANY TWO POINTS in a circuit,

$$I = \frac{\Delta V}{R}.$$

A central implication of this relation (Ohm's Law) is this:

$$I \propto \frac{1}{R}$$

Given a choice between two current branches,
the portion (fraction) of current that 'chooses' one particular branch is necessarily
INVERSELY proportional to the the fraction of total resistance found in that branch.

Mathematically, treatment of an inverse proportionality might seem 'obvious', but remember:
... You can rely on the seemingly simple pattern 'one goes up, the other goes down'
IFF the third term of the equation – – the one here referring to potential drop – – is a CONSTANT.
ΔV must be one single value applicable to both branches, NOT, for example,
some kind of 'total value that 'gets split' between the two!

Indeed, for any two paths or devices in parallel, the potential difference
is necessarily a constant . But this somewhat surprising (hard to remember) idea
comes from physics, not from math:

EVERY CHARGED PARTICLE - no matter which way it goes -
and therefore EVERY INDEPENDENT PATH that any charge might possibly follow
is independently subject to the laws of physics.
EVERY POSSIBLE PATH must therefore INDEPENDENTLY
OBEY ENERGY CONSERVATION.

So, any and all ways to get from point A to point B
within the same circuit
must do the same work
and impose the same energy effect
on each Coulomb of charge:

Therefore,
The potential drops across any two 'parallel' (independent) routes through a circuit
Are necessarily identical.

It is crucial that you are comfortable with the reasoning behind this certainty – –
More often than not, a person who feels hopelessly stuck
in the middle of a circuit problem is forgetting this vital piece of information.
And more often than not, s/he is forgetting it
because s/he doesn't really believe it. . .

(continued on next page)..

So, . . .

$I \alpha \frac{1}{R}$

This ~ 12.3 milliAmperes of current splits into three different portions,

but, much like considering forces when applying Newton's 2nd Law,
it is far clearer and more constructive to think of things
'like a charge': ONE CHOICE AT A TIME.
That is, each charge FIRST faces a split between Way #1 (through R6) or the remaining set of Ways: #2&#3, . . .
THEN a split between Way #2 (through R7) or Way #3 (through R8)

In other words, there really is no meaning or utility to any notion of a '3 – Way Choice'.
That's not a choice. That's a mess.

MUCH like this paragraph
might cease to be... if you are willing to slog through a bit more and ...
continue reading on the next page . . .

SO: . . .
R5 (500Ω) is in series with the battery: ALL 12.3 mA come through this resistor.

Then this 'Main Loop' of current (or 'Battery Current')
(two equally acceptable terms for, essentially, the trunk of a tree)
splits into a total of $\boxed{973}$ possible Ohms:
a 600 Ω path vs a 373 Ω path
$because \left(\frac{1}{700}+\frac{1}{800}\right)^{-1} \approx 373.$

So, $\frac{373}{973}$ of 12.3 mA flows through the the 600Ω ≈ 4.72 mA
and $\frac{600}{973}$ of 12.3 mA flows through the remaining split (700 Ω / 800 Ω) ≈ 7.58mA.

Then, between that 700 Ω and the 800 Ω (a total of 1500 Ω):

$\frac{800}{1500}$ of mA flows through the 700 Ohm ≈ 4.04 mA
and $\frac{700}{1500}$ of 2.41 mA flows through the 800 Ohm ≈ 3.54 mA.

*In conclusion:*

| |
|---|
| *i.* $R_5 (500\,\Omega)$: $I_5 \approx 12.3\ mA.$ |
| *ii.* $R_6 (600\Omega)$: $I_6 \approx 4.72\ mA.$ |
| *iii.* $R_7 (700\Omega)$: $I_7 \approx 4.04\ mA.$ |
| *iv.* $R_8 (800\Omega)$: $I_8 \approx 3.54\ mA.$ |

b) The ***potential difference*** across each and every RESISTOR (3 pts each, +1).

i. ***ΔV*** (R5) =

$\Delta V = IR.$

*So...*

$\Delta V_5 = I_5 R_5, etc.$

$\Delta V_5 = I_5 R_5$

$\approx (12.3\, mA)\,(500\, \Omega)...$

$\boxed{\Delta V_1 \approx 6.15\, Volts.}$

From Energy Conservation,

It SHOULD be the case that:

$\Delta V_3 \approx \Delta V_2 (\approx \Delta V_2) \approx 2.85\, Volts...$

*Let's Check:*

$\Delta V_2 \approx (4.72\, mA)\,(600\Omega)$

*ii.* $\boxed{\Delta V_2 \approx 2.83\, Volts \checkmark !}$

*Check:*

$\Delta V_3 \approx (4.04\, mA)\,(700\ \Omega)$

*iii.* $\boxed{\Delta V_3 \approx 2.83 Volts\ \checkmark !}$

*And, again, this time from the definition of*

*'parallel configuration',*

*It SHOULD be the case that*

$\Delta V_4 \approx \Delta V_3 \approx \Delta V_2 \approx 1.6 mA...$

*Check:*

$\Delta V_4 \approx (3.54\, mA)\,(800\, \Omega)$

*iv.* $\boxed{\Delta V_4 \approx 2.83\, Volts \checkmark !}$

IV. THE
B-FIELD (15 pts)
DIRECTION

Let:

$$\vec{A} \equiv 3\hat{x} - 5\hat{y} + \hat{z}$$
$$\vec{B} \equiv 6\hat{x} + 5\hat{y} + \hat{z}$$
$$\vec{C} \equiv 4$$

A. Find $\left(\vec{\mathbf{B}} \cdot \vec{\mathbf{A}}\right)$

B. Find $\vec{\mathbf{B}} \times \left(\vec{\mathbf{B}} \times \vec{\mathbf{A}}\right)$

C. Find $c\left(\vec{\mathbf{A}} \times \vec{\mathbf{B}}\right)$

Let the magnetic field produced by an infinitesimal amount of current
be given by the Biot-Savart Law:

$$d\vec{\mathbf{B}} = \frac{\mu_0}{4\pi}\frac{I d\vec{l}}{r^2} \times \hat{r}$$

and

Let the magnetic force exerted on a correspondingly small amount of current
by the magnetic field (produced in the manner described above) be

$$\vec{\mathbf{F}}_B = I d\vec{l} \times \vec{\mathbf{B}}$$

and let:

The direction of the vector produced by the 'cross' multiplication of two vectors,
If not otherwise adjudicated, be indicated by

***The 'Right-Hand' Rule.***

**Then:**

**D) Underneath a long straight current flowing to the right, in what direction does the magnetic field point?**

**E) If a rightward-flowing current is placed into a magnetic field which itself is pointing into the board, in what direction will the current be *forced* ?**

**F) A metal rod faces North/South on two metal rails that run East/West. There is a resistor running North/South and clamped down to complete a metal rectangle. Some sort of mechanical force is applied to the metal rod so that it is pulled Eastward along the rails and thereby progressively expands the area of the rectangle. The whole apparatus is submerged in a constant magnetic field which points up toward the ceiling.**

**1) In what direction will positive charges in the rod be forced to move (North? West?)**

**2) In what direction will current necessarily begin to flow? (clockwise? counterclockwise?)**

P204.S17.Final Exam
Problem 4 (abc). updated 5 – 29, 9'45am

$\vec{A} \equiv 3\hat{x} - 5\hat{y} + \hat{z}$
$\vec{B} \equiv 6\hat{x} + 5\hat{y} + \hat{z}$
$\vec{C} \equiv 4$

A) Find $\vec{B} \cdot \vec{A}$
$\vec{B} \cdot \vec{A} \equiv (B_x A_x) + (B_y A_y) + (B_z A_z)$
$\vec{B} \cdot \vec{A} = 18 + (-25) + 1$

$\boxed{\vec{B} \cdot \vec{A} = -6}$

B) Find $\vec{B} \times (\vec{B} \times \vec{A})$
$(\vec{B} \times \vec{A}) \equiv (B_y A_z - B_z A_y)\hat{x} + (B_z A_x - B_x A_z)\hat{y} + (B_x A_y - B_y A_x)\hat{z}$
$(\vec{B} \times \vec{A}) = [(5)(1) - (1)(-5)]\hat{x} + [(1)(3) - (6)(1)]\hat{y} + [(6)(-5) - (5)(3)]\hat{z}$

$\boxed{(\vec{B} \times \vec{A}) = 10\hat{x} - 3\hat{y} - 45\hat{z}}$

C) Find $c(\vec{A} \times \vec{B})$
If $(\vec{B} \times \vec{A}) = 10\hat{x} - 3\hat{y} - 45\hat{z}$
then $(\vec{A} \times \vec{B}) \equiv -(\vec{B} \times \vec{A}) = -10\hat{x} + 3\hat{y} + 45\hat{z}$
and then $c(\vec{A} \times \vec{B}) = 4(-10\hat{x} + 3\hat{y} + 45\hat{z})$

$\boxed{c(\vec{A} \times \vec{B}) = -40\hat{x} + 12\hat{y} + 180\hat{z}}$

D) Into the 'board'; into the plane of the current and point of interest

E) UP

F)

1. SOUTH

2. CLOCKWISE

# PART III

SUMMER 2014

# 5
# FULL-TEXT OF FINAL EXAM SU14: BLANK

# *Final EXAM:*

# E, M & Radiation

**PHYSICS 204**

**DANIEL A. MARTENS YAVERBAUM**

**JOHN JAY COLLEGE OF CRIMINAL JUSTICE, THE CUNY**

**JULY 26, 2014**

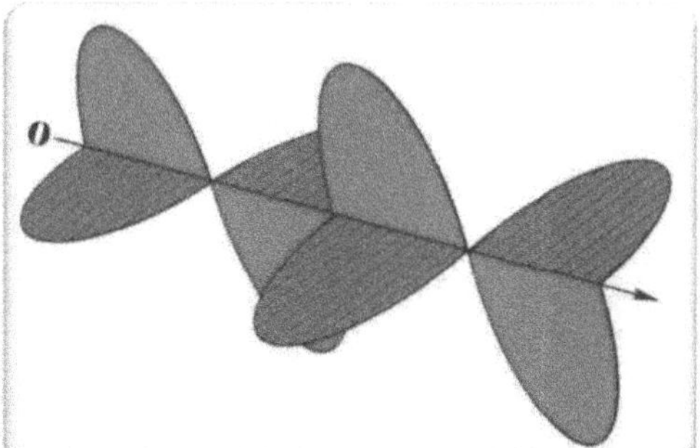


**Name:** ______________________________

**Section #:** ____________

SCORE: __________

## SOME USEFUL RELATIONS:

1) $\oint \vec{E} \cdot \overrightarrow{da} = \frac{q_{(enc)}}{\varepsilon_0}$.

2) $\oint \vec{B} \cdot \overrightarrow{da} = 0$.

3) $\oint \vec{E} \cdot \overrightarrow{dl} = -\frac{d}{dt} \int \vec{B} \cdot \overrightarrow{da}$.

4) $\oint \vec{B} \cdot \overrightarrow{dl} = \mu_0 I_{(enc)} + \mu_0 \varepsilon_0 \frac{d}{dt} \int \vec{E} \cdot \overrightarrow{da}$.

5) $\vec{E} = \frac{1}{4\pi\varepsilon_0} \frac{q}{r^2} \hat{r}$.

6) $\vec{F}_E \equiv q\vec{E}$.

7) $\vec{B} = \frac{\mu_0}{4\pi} \frac{q}{r^2} \vec{v} \times \hat{r}$.

8) $\vec{F}_B \equiv q\vec{v} \times \vec{B} = \vec{I}l \times \vec{B}$.

9) $V_a - V_b \equiv \int_a^b \vec{E} \cdot \overrightarrow{dr}$.

10) $C \equiv \frac{Q}{\Delta V}$. (Note: This "C" is capital.)

11) $I \equiv \frac{dq}{dt}$

12) $I = \frac{\Delta V}{R}$.

13) $\mathcal{E} - IR - \frac{Q}{C} = 0$. (Capital "C".)

14) $c \approx 3 \; x \; 10^8 \; m/s$.

15) $n \equiv \frac{c}{v}$. (Lower-case "c".)

16) $n_1 \sin\theta_1 = n_2 \sin\theta_2$.

17) $\frac{Yd}{L} = \frac{n\lambda}{2}$

18) $sin^2\theta + \; cos^2\theta = 1$.

SCORE: ______________

19) $\bar{v} \equiv \frac{\Delta x}{\Delta t}$.

20) $\Delta\, t' = \frac{\Delta t}{\sqrt{1-\frac{v^2}{c^2}}}$. Lower-case "c". Bold-face conclusion.

21) $\sum \vec{F} = m\vec{a}$.

22) $F = -Kx$.

23) $x = Acos(\omega t)$.

24) $\omega = 2\pi f$.

25) $f = \frac{1}{T}$.

26) $KE = \frac{1}{2}mv^2$

27) $PE_{elastic} = \frac{1}{2}Kx^2$

28) $v = \lambda f$.

29) $v = \sqrt{\frac{T}{\mu}}$.

30) $\frac{\partial^2 y}{\partial t^2} = (v^2)\frac{\partial^2 y}{\partial x^2}$.

31) $L = \frac{n\lambda}{2}$.

32) $\Delta l = \frac{n\lambda}{2}$.

34) $\epsilon_0 \approx 8.85\ x\ 10^{-12}\ \frac{C^2}{Nm^2}$.

35) $\mu_0 \approx 1.26 \times 10^{-7}\ \frac{N}{A^2}$.

36) $m_e \approx 9.11\ x\ 10^{-31}$ kg.

37) $q_e \approx 1.60\ x\ 10^{-19}\ C$.

SCORE: _______________

# I. Electrostatics (from a continuous charge distribution) (45 pts).

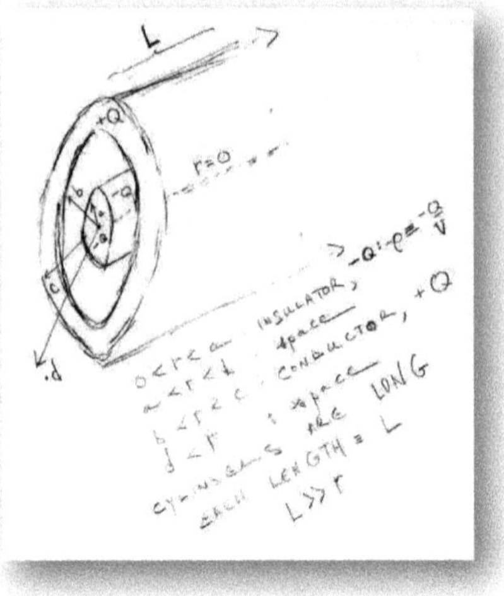


The above depicts the face-on view (cross-section) of a set of concentric ***CYLINDERS***, EACH of LENGTH ***L***.

The innermost cylinder, radius ***a***, is made of ***insulating*** material.
This ***insulator*** contains a net ***negative*** charge of **-*Q***.
The net ***negative*** charge, **-*Q***, is uniformly distributed throughout the ***cylinder***; $-\rho \equiv \frac{-Q}{V}$.

Outside the ***insulating cylinder***, there is some empty space (vacuum):
Here, there is simply no material to contain charges of any kind.
The space extends from radius ***a*** to a larger radius, ***b***.

Both the ***insulating cylinder*** and the space are surrounded by a ***cylindrical shell***—inner radius, ***b***, outer radius **c**.

This ***cylindrical shell*** is made of ***conducting material***.
The conducting ***shell*** contains a net ***positive*** charge of +***Q***.

Point ***d*** refers to an arbitrary location of interest fully outside the entire arrangement configuration of cylinder within shell.

This entire configuration has been sitting on a lab table for a long time; it is electrically stable.

SCORE: ________________

a) Quickly sketch a simplified version of the diagram above. Then draw +'s and –'s to ***indicate where*** all ***charges are*** located (2 pts).

b) Use your simplified sketch above. ***Draw*** properly representative ***field lines*** anywhere they apply. Draw enough lines to provide a clear sense of ***direction and*** comparative ***magnitude*** (3 pts).

c) Determine the magnitude of the ***electric field*** as a function of position (measured out from the insulator's central axis) for each of the following regions:

i. b < r < c: $E(r) = ?$ (2 pts.)

ii. a < r < b: $E(r) = ?$ (5 pts.)

SCORE: ______________

iii. $0 < r < a$: $\boldsymbol{E}(r) = ?$ (3 pts.)

iv. $d < r$: $\boldsymbol{E}(r) = ?$ (2 pts.)

d) Determine the ***electric potential difference*** as a function of position across each of the following regions. If it helps you, you may assume that $V \to 0$ *as* $r \to \infty$.

i. $V(a) - V(0) = ?$ (5 pts.)

ii. $V(0) - V(b) = ?$ (5 pts.)

SCORE: ______________

iii. $V(b) - V(c) = ?$ (2 pts.)

iv. $V(o) - V(d) = ?$ (2 pts.)

v. $V(d) - V(o) = ?$ (2 pts.)

vi. $V(o) - V(\infty) = ?$ (2 pts.)

e) Find ***C***, the ***capacitance*** of this physical arrangement:
Specifically, find ***C*** between ***r* = *o*** and ***r* = *d*** (3 pts).

(We say ***d***, rather than **c**, just to make absolutely certain that we account for every possible bit of material, charge and space included in this arrangement.)

Your answer will be expressed in terms of given and fundamental constants.

SCORE: ______________

Now assume the following:

***a* = 5 cm.**
***b* = 10 cm.**
**c = 8 cm.**
***L* = 300 cm.**

The center of the ***insulating cylinder*** is connected by wire to a ***150 ohm resistor***. The other end of this resistor is connected to the negative terminal of a ***battery***. The positive terminal of the battery is connected to the ***conducting shell***.

f) Draw a circuit diagram for this new situation. You may use the standard symbol for capacitor even though this capacitor happens not to be composed of plates (2 pts).

g) From the moment the wires are all connected, current flows. As it flows, its rate of flow continuously slows down.

From the moment the wires are all connected, how much time will pass until this current decays to 1 ***pi***th (approx.. $\frac{1}{3.14...}$) of its initial value (5 pts)?

THIS ANSWER WILL BE NUMERICAL – and should be expressed with appropriate units.

SCORE: ______________

## II. A CIRCUIT (20 pts).

A. Examine the following two circuits (Values given under "Circuit Properties".)

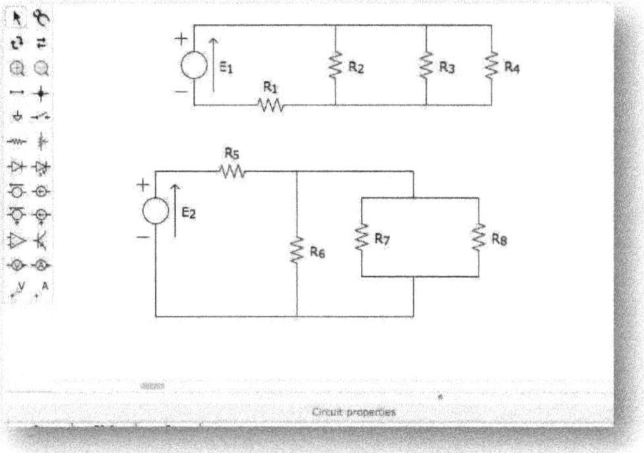


| | | |
|---|---|---|
| E1 | E2 | 15 Volts |
| R1 | R5 | 300 Ohms |
| R2 | R6 | 60 Ohms |
| R3 | R7 | 50 Ohms |
| R4 | R8 | 175 Ohms |

Is the ***potential difference*** across through R4

***Greater than that through*** $R_8$,
***Less than that through*** $R_8$
OR
***The same in both? In a complete English sentence or two, explain! (4 pts.)***

SCORE: ______________

B. For JUST THE FIRST of the two circuits, determine:

*i.* The ***current*** flowing through each and every RESISTOR (8 pts: 2 pts each).

I ($R_1$):

I ($R_2$):

I ($R_3$):

I ($R_4$):

SCORE: ______________

ii. The ***potential difference*** across each and every RESISTOR (8 pts: 2 pts each).

$\Delta V$ ($R_1$):

$\Delta V$ ($R_2$):

$\Delta V$ ($R_3$):

$\Delta V$ ($R_4$):

SCORE: _______________

## III. B-Fields (20 pts).

A. Use AMPERE's LAW in order to determine the ***magnetic field*** as a function of ***position***
near a LONG, STRAIGHT CURRENT-CARRYING WIRE, ***I*** (10 pts).

DRAW AN APPROPRIATE and fully labeled DIAGRAM!

That is, determine $\vec{B}(r)$.

B. Use the DIRECTIONS demanded by the ***Biot-Savart*** and ***Lorentz Force Laws*** to explain a historical and common every-day finding:
***WHY*** do two cylindrical bar magnets placed near each other in a certain orientation attract, yet repel once one magnet is rotated 180 degrees (10 pts)?

SCORE: ______________

## IV. Electromagnetic Radiation (15 pts).

FIND & CORRECT ***AT LEAST FOUR IMPORTANT MISTAKES***
IN THE FOLLOWING ARGUMENT. (There are more than 4.)

THE MORE SPECIFICALLY YOU CAN EXPLAIN THE MEANING and/or
SIGNIFICANCE OF THE MISTAKES, the more points you will gain!

*** The Argument →

According to Faraday's Law for Electromagnetic Induction:

$$\oint \vec{E} \cdot \overrightarrow{dl} = -\frac{d}{dt} \int \vec{B} \cdot \overrightarrow{da}.$$

One reasonable way to express the meaning of Faraday's Law is this:

An ***electric field***
will be induced through the ***closed path***
bounding some ***open area***
whenever the ***magnetic flux***
through that area
***changes*** as a ***function of time.***

With the inclusion of Maxwell's Displacement Current, the corrected and most general possible version of Ampere's Law becomes:

$$\oint \vec{B} \cdot \overrightarrow{dl} = \mu_0 I_{(enc)} + \mu_0 \varepsilon_0 \frac{d}{dt} \int \vec{E} \cdot \overrightarrow{da}.$$

One reasonable way to express the meaning of Maxwell's Corrected (version of) Ampere's Law is this:

a ***magnetic field***
will be induced through the ***closed area***
bounding some ***open volume***
whenever the ***electric flux***
through that area
***changes*** as a ***function of time***.

SCORE: ____________

We can then look at what must be true in the free space—far away from any pieces of charge or current. There, the four Maxwell's Equations become:

$$1)\ \oint \vec{E} \cdot \vec{da} = 0.$$

$$2)\ \oint \vec{B} \cdot \vec{da} = 0.$$

$$3)\ \oint \vec{E} \cdot \vec{dl} = \frac{d}{dt}\int \vec{B} \cdot \vec{da}.$$

$$4)\ \oint \vec{B} \cdot \vec{dl} = \mu_0 \varepsilon_0 \frac{d}{dt}\int \vec{E} \cdot \vec{da}.$$

Rearranging ("decoupling") the equations so that we can look at electric fields on their own and magnetic fields on their own,
and assuming that *A* stands for some sort of space variable
(like x, but more general, so that it can include 2 or 3-dimensional space coordinates),

We find:

$$\frac{\partial^2}{\partial A^2}\vec{E} = \mu_0 \varepsilon_0 \frac{\partial^2}{\partial t^2}\vec{B}$$

and

$$\frac{\partial^2}{\partial A^2}\vec{B} = \mu_0 \varepsilon_0 \frac{\partial^2}{\partial t^2}\vec{E}.$$

That is, we find:

Any Current that oscillates harmonically
(like $I = I_0 \cos(\omega t)$)
will create a Electric Field that oscillates harmonically
(like $E = -E_0 \sin \omega t$)).
Any Electric Field that oscillates harmonically
will induce an Magnetic Field that oscillates harmonically
(like $B = -B\cos(\omega t)$)
which will induce a Electric Field that oscillates harmonically…

SCORE: ______________

And the oscillating Electric and Magnetic Fields
will continue to
induce each other,
each field making
a 90 degree
right-hand turn from the other

so that these patterns of oscillations
necessarily travel through space
out and away from any source charge or source current.

As stated above, these traveling field oscillations perpetually obey

$$\frac{\partial^2}{\partial A^2}\vec{E} = \mu_0\varepsilon_0\frac{\partial^2}{\partial t^2}\vec{E}$$

and

$$\frac{\partial^2}{\partial A^2}\vec{B} = \mu_0\varepsilon_0\frac{\partial^2}{\partial t^2}\vec{B};$$

The fields are therefore solutions to equations of the same form as:

$$\frac{\partial^2}{\partial t^2}y = v^2\frac{\partial^2}{\partial y^2}x$$

The mutually inducing perpendicular fields therefore
travel through space as three-dimensional WAVES for which

$$v = \frac{1}{\sqrt{\mu_0\varepsilon_0}} = \boldsymbol{c}$$

Since the medium for these waves are fields,
And since all fields extend infinitely through all space,
these waves travel at a speed
which is constant
relative to
all of space.

Thus, the head-***light*** from an express train travels at ***c***
relative to observers on that speeding train, yet also
travels at (that same) ***c***
relative to observers on the platform.

(And this, therefore, is either the beginning of the end or the end of the beginning.)

SCORE: _______________

# 6
# FULL-TEXT OF FINAL EXAM SU14: SOLVED & EXPLAINED

# Final EXAM:

# E, M & Radiation

## PHYSICS 204

DANIEL A. MARTENS YAVERBAUM

JOHN JAY COLLEGE OF CRIMINAL JUSTICE, THE CUNY

July 26, 2014

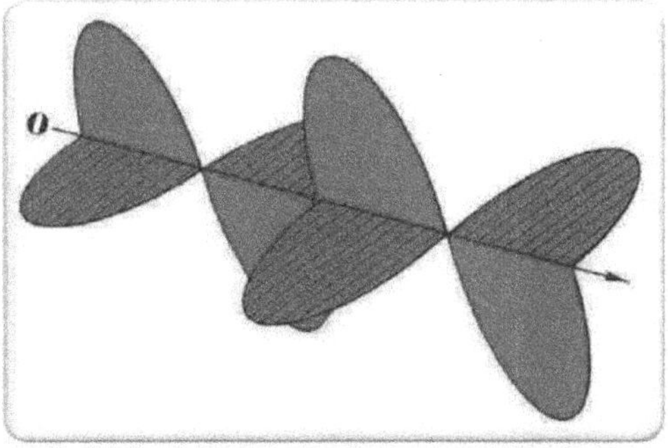


Name: SOLUTIONS!

Section #: ____________

SCORE: ____________

## SOME USEFUL RELATIONS:

1) $\oint \vec{E} \cdot \overrightarrow{da} = \frac{q_{(enc)}}{\varepsilon_0}$.

2) $\oint \vec{B} \cdot \overrightarrow{da} = 0$.

3) $\oint \vec{E} \cdot \overrightarrow{dl} = -\frac{d}{dt} \int \vec{B} \cdot \overrightarrow{da}$.

4) $\oint \vec{B} \cdot \overrightarrow{dl} = \mu_0 I_{(enc)} + \mu_0 \varepsilon_0 \frac{d}{dt} \int \vec{E} \cdot \overrightarrow{da}$.

5) $\vec{E} = \frac{1}{4\pi\varepsilon_0} \frac{q}{r^2} \hat{r}$.

6) $\vec{F}_E \equiv q\vec{E}$.

7) $\vec{B} = \frac{\mu_0}{4\pi} \frac{q}{r^2} \vec{v} \times \hat{r}$.

8) $\vec{F}_B \equiv q\vec{v} \times \vec{B} = \vec{I}l \times \vec{B}$.

9) $V_a - V_b \equiv \int_a^b \vec{E} \cdot \overrightarrow{dr}$.

10) $C \equiv \frac{Q}{\Delta V}$. (Note: This "C" is capital.)

11) $I \equiv \frac{dq}{dt}$

12) $I = \frac{\Delta V}{R}$.

13) $\mathcal{E} - IR - \frac{Q}{C} = 0$. (Capital "C".)

14) $c \approx 3 \, x \, 10^8 \, m/s$.

15) $n \equiv \frac{c}{v}$. (Lower-case "c".)

16) $n_1 \sin \theta_1 = n_2 \sin \theta_2$.

17) $\frac{Yd}{L} = \frac{n\lambda}{2}$

18) $sin^2\theta + cos^2\theta = 1$.

SCORE: ______________

19) $\bar{v} \equiv \frac{\Delta x}{\Delta t}$.

20) $\Delta t' = \frac{\Delta t}{\sqrt{1-\frac{v^2}{c^2}}}$. Lower-case "c". Bold-face conclusion.

21) $\sum \vec{F} = m\vec{a}$.

22) $F = -Kx$.

23) $x = Acos(\omega t)$.

24) $\omega = 2\pi f$.

25) $f = \frac{1}{T}$.

26) $KE = \frac{1}{2}mv^2$

27) $PE_{elastic} = \frac{1}{2}Kx^2$

28) $v = \lambda f$.

29) $v = \sqrt{\frac{T}{\mu}}$.

30) $\frac{\partial^2 y}{\partial t^2} = (v^2)\frac{\partial^2 y}{\partial x^2}$.

31) $L = \frac{n\lambda}{2}$.

32) $\Delta l = \frac{n\lambda}{2}$.

34) $\epsilon_0 \approx 8.85 \, x \, 10^{-12} \, \frac{C^2}{Nm^2}$.

35) $\mu_0 \approx 1.26 \times 10^{-7} \, \frac{N}{A^2}$.

36) $m_e \approx 9.11 \, x \, 10^{-31}$ kg.

37) $q_e \approx 1.60 \, x \, 10^{-19} \, C$.

SCORE: ______________

# I. Electrostatics (from a continuous charge distribution) (45 pts).

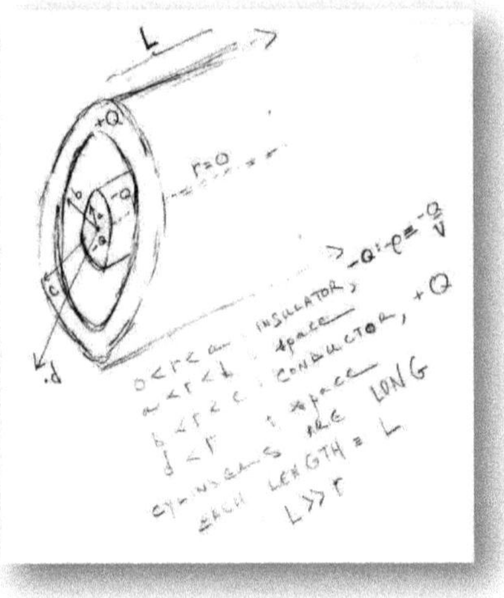


The above depicts the face-on view (cross-section) of a set of concentric ***CYLINDERS***, EACH of LENGTH ***L.***

The innermost cylinder, radius ***a***, is made of ***insulating*** material.
This ***insulator*** contains a net ***negative*** charge of **-*Q*.**
The net ***negative*** charge, **-*Q***, is uniformly distributed throughout the ***cylinder***; $-\rho \equiv \frac{-Q}{V}$.

Outside the ***insulating cylinder***, there is some empty space (vacuum):
Here, there is simply no material to contain charges of any kind.
The space extends from radius ***a*** to a larger radius, ***b***.

Both the ***insulating cylinder*** and the space are surrounded by a ***cylindrical shell***— inner radius, ***b***, outer radius **c**.

This ***cylindrical shell*** is made of ***conducting material***.
The conducting ***shell*** contains a net ***positive*** charge of +***Q*.**

Point ***d*** refers to an arbitrary location of interest fully outside the entire arrangement configuration of cylinder within shell.

This entire configuration has been sitting on a lab table for a long time; it is electrically stable.

SCORE: ________________

a) Quickly sketch a simplified version of the diagram above. Then draw +'s and –'s to ***indicate where*** all ***charges are*** located (2 pts).

b) Use your simplified sketch above. ***Draw*** properly representative ***field lines*** anywhere they apply. Draw enough lines to provide a clear sense of ***direction and*** comparative ***magnitude*** (3 pts).

c) Determine the magnitude of the ***electric field*** as a function of position (measured out from the insulator's central axis) for each of the following regions:

i. b < r < c: $E(r) = ?$ (5 pts.)

$$E(r) \equiv \mathbf{0} \text{ N/C}$$

ii. a < r < b: $E(r) = ?$ (5 pts.)

$$\oint\!\!\oint \mathbf{E} \cdot \vec{\mathbf{ds}} = \frac{q_{(encl)}}{\epsilon_0}$$

$$E \oint\!\!\oint ds = \frac{q_{(encl)}}{\epsilon_0}$$

$$E \cdot s = \frac{q_{(encl)}}{\epsilon_0}$$

$$E \cdot \pi r L = \frac{q_{(encl)}}{\epsilon_0}$$

$$E \cdot 2\pi r \cdot L = \frac{Q}{\epsilon_0}$$

$$E = \frac{1}{2\pi\epsilon_0} \cdot \frac{Q}{Lr}$$

$$\boxed{\begin{aligned} E &\equiv \|\vec{\mathbf{E}}\| \\ &= \left(\frac{1}{2\pi\epsilon_0}\right)\frac{\lambda}{r} \end{aligned}}$$

SCORE: ______________

iii. $0 < r < a$: $E(r) = ?$ (3 pts.)

$$\oiint \vec{E} \cdot d\vec{s} = \frac{q_{(enc)}}{\epsilon_0}$$
$$E \oiint ds = \frac{q_{(enc)}}{\epsilon_0}$$
$$E \cdot s = \frac{q_{(enc)}}{\epsilon_0}$$
$$E \cdot \pi r L = \frac{q_{(enc)}}{\epsilon_0}$$

Let $\tau \equiv$ volume. Then

$$\frac{q_{(enc)}}{\tau_{(enc)}} = \frac{Q_{(total)}}{\tau_{(total)}}$$
$$\frac{q_{(enc)}}{\pi r^2 L} = \frac{Q_{(total)}}{\pi a^2 L}$$
$$q_{(enc)} = \frac{Q_{(total)} \, \pi r^2 L}{\pi a^2 L}$$
$$q_{(enc)} = Q \frac{r^2}{a^2}$$

iv. $d < r$: $E(r) = ?$ (2 pts.)

$E(r) \equiv 0$ N/C

and:

$$E \cdot \pi r L = Q \frac{r^2}{\epsilon_0 a^2}$$

$$\boxed{E = \frac{Q}{2\pi\epsilon_0} \cdot \frac{r^2}{a^2 L}}$$

d) Determine the ***electric potential difference*** as a function of position across each of the following regions. If it helps you, you may assume that $V \to 0$ *as* $r \to \infty$.

i. $V(a) - V(0) = ?$ (5 pts.)

$$\Delta V \equiv V_0 - V_a$$
$$\equiv \int_{r=0}^{r=a} \vec{E} \cdot \vec{dr}$$
$$= \int_{r=0}^{r=a} \left( \frac{1}{2\pi\epsilon_0} \cdot \frac{Qr}{La^2} \right) dr$$
$$= \frac{1}{2\pi\epsilon_0} \left( \frac{Q}{La^2} \right) \int_{r=0}^{r=a} r \, dr$$
$$= \left( \frac{1}{2\pi\epsilon_0} \right) \left( \frac{Q}{La^2} \right) r \Big|_{r=0}^{r=a}$$
$$= \left( \frac{1}{2\pi\epsilon_0} \right) \left( \frac{Q}{La^2} \right) \frac{a^2}{2}$$

$$\boxed{\Delta V \equiv V_0 - V_a = \left( \frac{1}{4\pi\epsilon_0} \right) \left( \frac{Q}{L} \right)}$$

ii. $V(0) - (b) = ?$ (5 pts.)

$$V_a - V_b \equiv \int_{r=a}^{r=b} \vec{E} \cdot \vec{dr}$$
$$= \int_{r=a}^{r=b} \left( \frac{1}{2\pi\epsilon_0} \cdot \frac{Q}{Lr} \right) dr$$
$$= \frac{1}{2\pi\epsilon_0} \left( \frac{Q}{L} \right) \int_{r=a}^{r=b} \frac{1}{r} dr$$
$$= \left( \frac{1}{2\pi\epsilon_0} \right) \left( \frac{Q}{L} \right) \ln r \Big|_{r=a}^{r=b}$$
$$= \left( \frac{1}{2\pi\epsilon_0} \right) \left( \frac{Q}{L} \right) (\ln b - \ln a)$$
$$= \left( \frac{1}{2\pi\epsilon_0} \right) \left( \frac{Q}{L} \right) \left( \ln \frac{b}{a} \right)$$

$$\Delta V \equiv (V_a - V_b) + (V_0 - V_a)$$
$$= \left( \frac{1}{2\pi\epsilon_0} \right) \left( \frac{Q}{L} \right) \left( \ln \frac{b}{a} \right) + \left( \frac{1}{4\pi\epsilon_0} \right) \frac{Q}{L}$$

$$\boxed{V_0 - V_b = \left( \frac{1}{2\pi\epsilon_0} \right) \left( \frac{Q}{L} \right) \left[ \ln \frac{b}{a} + \frac{1}{2} \right]}$$

SCORE: ____________

iii. V(b) – (c) = ? (3 pts.)

$$\Delta V \equiv 0 \text{ Volts}$$

iv. V(c) – V(d) = ? (2 pts.)

$$\Delta V \equiv 0 \text{ Volts}$$

v. V(0) – V(d) = ? (3 pts.)

$$V_0 - V_b = \left(\frac{1}{2\pi\epsilon_0}\right)\left(\frac{Q}{L}\right)\left[\ln\frac{b}{a} + \frac{1}{2}\right]$$

vi. V(0) – $V(\infty)$ = ? (2 pts.)

$$V_0 - V_b = \left(\frac{1}{2\pi\epsilon_0}\right)\left(\frac{Q}{L}\right)\left[\ln\frac{b}{a} + \frac{1}{2}\right]$$

**e) Find *C*, the *capacitance* of this physical arrangement:**
**Specifically, find *C* between $r = o$ and $r = d$ (3 pts).**

**(We say *d*, rather than *c*, just to make absolutely certain that we account for every possible bit of material, charge and space included in this arrangement.)**

**Your answer will be expressed in terms of given and fundamental constants.**

$$C \equiv \frac{q}{\Delta V}$$

$$C = \frac{Q}{\left(\frac{1}{2\pi\epsilon_0}\right)\left(\frac{Q}{L}\right)\left[\ln\frac{b}{a} + \frac{1}{2}\right]}$$

$$C = \frac{2QL\pi\epsilon_0}{Q\left[\ln\frac{b}{a} + \frac{1}{2}\right]}$$

$$C = \frac{2\pi\epsilon_0 L}{\left[\ln\frac{b}{a} + \frac{1}{2}\right]}$$

SCORE: ______________

Now assume the following:

***a* = *5 cm.***
***b* = *10 cm.***
**c = *8 cm.***
***L* = *300 cm.***

The center of the ***insulating cylinder*** is connected by wire to a ***150 ohm resistor***.
The other end of this resistor is connected to the negative terminal of a ***battery***.
The positive terminal of the battery is connected to the ***conducting shell***.

f) Draw a circuit diagram for this new situation. You may use the standard symbol for capacitor even though this capacitor happens not to be composed of plates (2 pts).

g) From the moment the wires are all connected, current flows. As it flows, its rate of flow continuously slows down.

From the moment the wires are all connected, how much time will pass until this current decays to 1 ***pi***$^{th}$ (approx.. $\frac{1}{3.14...}$) of its initial value (5 pts)?

THIS ANSWER WILL BE NUMERICAL – and should be expressed with appropriate units.

SCORE: ______________

**Find t such that:**

$$I = \frac{I_0}{\pi}$$

$$\varepsilon - IR - \frac{q}{C} = 0$$

$$\varepsilon - \frac{dq}{dt}R - \frac{q}{C} = 0$$

$$\varepsilon - \frac{q}{C} = \frac{dq}{dt}R$$

$$\frac{dt}{R} = \frac{dq}{\varepsilon - \frac{q}{C}}$$

$$\frac{1}{R}dt = \frac{1}{\varepsilon - \frac{q}{C}}dq$$

$$\frac{1}{R}\int dt = \int \frac{dq}{\varepsilon - \frac{q}{C}}$$

$$\frac{t}{R} + x = \int \frac{dq}{\varepsilon - \frac{q}{C}}$$

$$\text{let } u \equiv \varepsilon - \frac{q}{C}$$

$$\text{then } \frac{du}{dq} = -\frac{1}{C}$$

$$du = -\frac{dq}{C}$$

$$dq = -C\,du$$

**So:**

$$\frac{t}{R} + x = \int \frac{-C\,du}{u}$$

$$\frac{t}{R} + x = -C\int \frac{du}{u}$$

$$\frac{t}{R} + x = -C\ln\left|\varepsilon - \frac{q}{C}\right|$$

$$-\frac{t}{RC} + x = \ln\left|\varepsilon - \frac{q}{C}\right|$$

$$e^{-\frac{t}{RC}}e^{x} = \varepsilon - \frac{q}{C}$$

$$xe^{-\frac{t}{RC}} = \varepsilon - \frac{q}{C}$$

$$\frac{q}{C} = \varepsilon - xe^{-\frac{t}{RC}}$$

$$q = \varepsilon C - xe^{-\frac{t}{RC}}$$

**One arbitrary constant remains:**

$$x = ?$$

**Consider initial condition:**

$$\text{at } t = 0:$$

$$q \equiv q_0 = 0$$

**so:**

$$0 = \varepsilon C - x$$

$$x = \varepsilon C$$

**Thus**

$$q = \varepsilon C - \varepsilon C\left(e^{-\frac{t}{RC}}\right)$$

$$I \equiv \frac{dq}{dt} = 0 \not- \varepsilon C\left(e^{-\frac{t}{RC}}\right)\left(\not-\frac{1}{RC}\right)$$

$$I = \frac{\varepsilon}{R}e^{-\frac{t}{RC}}$$

$$I_0 \equiv I_{(t=0)} = \frac{\varepsilon}{R}e^{0}$$

$$I_0 = \frac{\varepsilon}{R}$$

**We seek:** $I = \frac{I_0}{\pi} \Rightarrow$

$$\left(\frac{\varepsilon}{R}\right)e^{-\frac{t}{RC}} = \left(\frac{\varepsilon}{R}\right)\frac{1}{\pi}$$

$$-\frac{t}{RC} = \ln\left(\frac{1}{\pi}\right)$$

$$t = -RC\ln\left(\frac{1}{\pi}\right)$$

$$t = RC\ln\pi$$

$$t \approx (150)(1.40 \times 10^{-10})\ln\pi$$

$$\boxed{t \approx 2.40 \times 10^{-8}\,\text{sec}}$$

II. A CIRCUIT (20 pts).

A. Examine the following two circuits (Values given under "Circuit Properties".)

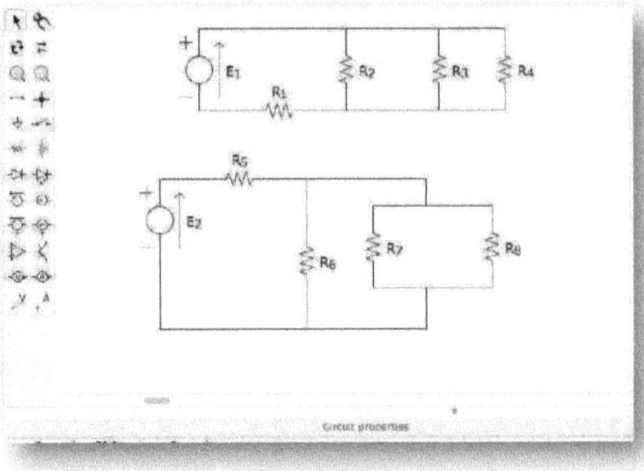


| | | |
|---|---|---|
| E1 | E2 | 15 Volts |
| R1 | R5 | 300 Ohms |
| R2 | R6 | 60 Ohms |
| R3 | R7 | 50 Ohms |
| R4 | R8 | 175 Ohms |

Is the ***potential difference*** across through R4

***Greater than that through $R_8$,***
***Less than that through $R_8$***
OR
***The same in both? In a complete English sentence or two, explain! (4 pts.)***

SCORE: ______________

B. For JUST THE FIRST of the two circuits, determine:

*i.* The ***current*** flowing through each and every RESISTOR (8 pts: 2 pts each).

I (R1):

B) Current.

i. Through $R_1$:

$$\frac{1}{R_{45}} = \frac{1}{R_4} + \frac{1}{R_5}$$

$$\frac{1}{R_{45}} = \frac{1}{175} + \frac{1}{50}$$

$$\frac{1}{R_{45}} = \frac{2}{350} + \frac{7}{350}$$

$$\frac{1}{R_{45}} = \frac{9}{350}$$

$$R_{45} \approx 38.9\ \text{A}$$

$$\frac{1}{R_{452}} = \frac{1}{38.9} + \frac{1}{60}$$

$$\frac{1}{R_{452}} \approx .025 + .017$$

$$\frac{1}{R_{452}} \approx .042$$

$$R_{452} \approx 23.9\ \Omega$$

$$R_{1254} \approx 324\ \Omega$$

$$I_1 \approx \frac{\varepsilon}{R_{4521}}$$

$$I_1 \approx \frac{15\ \text{V}}{324\ \Omega}$$

$$\boxed{I_1 \approx .0462\ \text{A}}$$

$$\boxed{I_1 \approx 46.2\ \text{mA}}$$

I (R2):

$$R_1 = 300\ \Omega$$

$$\Delta V_1 \approx (.0462\ \text{A})(300\ \Omega)$$

$$\boxed{\Delta V_1 \approx 13.9\ \text{V}}$$

$$\Delta V_2 \approx 15 - 13.9$$

$$\boxed{\Delta V_2 = \Delta V_3 = \Delta V_4 \approx 1.1\ \text{V}}$$

$$I_2 = \frac{\Delta V_2}{R_2} \approx \frac{(1.1\text{V})}{(60\Omega)} \approx \frac{(1.1\text{V})}{(60\Omega)}$$

$$\boxed{I_2 \approx .0183\ \text{A}}$$

$$\boxed{I_2 \approx 18.3\ \text{mA}}$$

I (R3):

$$I_3 = \frac{\Delta V_3}{R_3} \approx \frac{(1.1\text{V})}{(50\Omega)}$$

$$\boxed{I_3 \approx .022\ \text{A}}$$

$$\boxed{I_3 \approx 22\ \text{mA}}$$

I (R4):

$$I_4 = \frac{\Delta V_4}{R_4} \approx \frac{(1.1\text{V})}{(175\Omega)}$$

$$\boxed{I_4 \approx .00629\ \text{A}}$$

$$\boxed{I_4 \approx 6.29\ \text{mA}}$$

SCORE: ____________

ii. The ***potential difference*** across each and every RESISTOR (8 pts: 2 pts each).

ΔV (R1):

$$I_1 \approx .0462\ A$$
$$I_1 \approx 46.2\ mA$$

$$R_1 = 300\ \Omega$$
$$\Delta V_1 \approx (.0462\ A)(300\ \Omega)$$

ΔV (R2):

$$\Delta V_1 \approx 13.9\ V$$

$$\Delta V_2 \approx 15 - 13.9$$

ΔV (R3):

$$\Delta V_2 = \Delta V_3 = \Delta V_4 \approx 1.1V$$

ΔV (R4):

SCORE: ____________

III. **B-Fields** (18 pts).

**IV.** **Use AMPERE's LAW in order to determine the *magnetic field* as a function of *position* near a LONG, STRAIGHT CURRENT-CARRYING WIRE, *I* (8 pts).**

DRAW AN APPROPRIATE and fully labeled DIAGRAM!

**That is, determine $\vec{B}(r)$.**

$$\oint \mathbf{B} \cdot d\mathbf{l} = \mu_0 I_{(enc)}$$
$$B \oint dl = \mu_0 I_{(enc)}$$
$$Bl = \mu_0 I_{(enc)}$$
$$B 2\pi r = \mu_0 I_{(enc)}$$
$$B 2\pi r = \mu_0 I$$

$$\boxed{B = \frac{\mu_0}{2\pi} \cdot \frac{I}{r}}$$

B. **Use the DIRECTIONS demanded by the *Biot-Savart* and *Lorentz Force Laws* to explain a historical and common every-day finding:**
*WHY* do two cylindrical bar magnets placed near each other in a certain orientation attract, yet repel once one magnet is rotated 180 degrees (10 pts)?

SEE SOLUTION TO

PROBLEM IV, Part I, of SU's Final Exam.

SCORE: ____________

## IV. Electromagnetic Radiation (12 pts).

FIND & CORRECT ***AT LEAST FOUR IMPORTANT MISTAKES (3 pts each)*** IN THE FOLLOWING ARGUMENT. (There are more than 4.)

THE MORE SPECIFICALLY YOU CAN EXPLAIN THE MEANING and/or SIGNIFICANCE OF THE MISTAKES, the more points you will gain!

*** The Argument →

According to Faraday's Law for Electromagnetic Induction:

$$\oint \vec{E} \cdot \overrightarrow{dl} = -\frac{d}{dt}\int \vec{B} \cdot \overrightarrow{da}.$$

One reasonable way to express the meaning of Faraday's Law is this:

An *electric field*
will be induced through the ***closed path***
bounding some ***open area***
whenever the ***magnetic flux***
through that area
***changes*** as a ***function of time.***

With the inclusion of Maxwell's Displacement Current, the corrected and most general possible version of Ampere's Law becomes:

$$\oint \vec{B} \cdot \overrightarrow{dl} = \mu_0 I_{(enc)} + \mu_0 \varepsilon_0 \frac{d}{dt}\int \vec{E} \cdot \overrightarrow{da}.$$

One reasonable way to express the meaning of Maxwell's Corrected (version of) Ampere's Law is this:

a ***magnetic field*** path
will be induced through the ***closed*** ~~***area***~~
bounding some ***open*** ~~***volume***~~
whenever the ***electric flux*** area
through that area
***changes*** as a ***function of time.***

SCORE: ______________

We can then look at what must be true in the free space—far away from any pieces of charge or current. There, the four Maxwell's Equations become:

$$1)\ \oint \vec{E} \cdot \overrightarrow{da} = 0.$$

missing the negative sign!

$$2)\ \oint \vec{B} \cdot \overrightarrow{da} = 0.$$

$$3)\ \oint \vec{E} \cdot \overrightarrow{dl} = \frac{d}{dt} \int \vec{B} \cdot \overrightarrow{da}.$$

$$4)\ \oint \vec{B} \cdot \overrightarrow{dl} = \mu_0 \varepsilon_0 \frac{d}{dt} \int \vec{E} \cdot \overrightarrow{da}.$$

Rearranging ("decoupling") the equations so that we can look at electric fields on their own and magnetic fields on their own,
and assuming that $A$ stands for some sort of space variable
(like x, but more general, so that it can include 2 or 3-dimensional space coordinates),

We find:

$$\frac{\partial^2}{\partial A^2} \vec{E} = \mu_0 \varepsilon_0 \frac{\partial^2}{\partial t^2} \vec{B}$$

and

$$\frac{\partial^2}{\partial A^2} \vec{B} = \mu_0 \varepsilon_0 \frac{\partial^2}{\partial t^2} \vec{E}.$$

That is, we find:

Any Current that oscillates harmonically
(like $I = I_0 \cos(\omega t)$)
will create a Electric Field that oscillates harmonically
(like $E = -E_0 \sin \omega t)$).
Any Electric Field that oscillates harmonically
will induce an Magnetic Field that oscillates harmonically
(like $B = -B\cos(\omega t)$)
which will induce a Electric Field that oscillates harmonically…

SCORE: ______________

And the oscillating Electric and Magnetic Fields
will continue to
induce each other,
each field making
a 90 degree
right-hand turn from the other

alternating:

left,
right,
left...

so that these patterns of oscillations
necessarily travel through space
out and away from any source charge or source current.

As stated above, these traveling field oscillations perpetually obey

$$\frac{\partial^2}{\partial A^2}\vec{E} = \mu_0\varepsilon_0\frac{\partial^2}{\partial t^2}\vec{E}$$

and

$$\frac{\partial^2}{\partial A^2}\vec{B} = \mu_0\varepsilon_0\frac{\partial^2}{\partial t^2}\vec{B};$$

The fields are therefore solutions to equations of the same form as:

$$\frac{\partial^2}{\partial t^2}y = v^2\frac{\partial^2}{\partial y^2}$$

$$\frac{d^2}{dt^2}y = v^2\frac{d^2}{dx^2}y$$

The mutually inducing perpendicular fields therefore
travel through space as three-dimensional WAVES for which

$$v = \frac{1}{\sqrt{\mu_0\varepsilon_0}} = c$$

Since the medium for these waves are fields,
And since all fields extend infinitely through all space,
these waves travel at a speed
which is constant
relative to
all of space.

Thus, the head-***light*** from an express train travels at $c$
relative to observers on that speeding train, yet also
travels at (that same) $c$
relative to observers on the platform.

(And this, therefore, is either the beginning of the end or the end of the beginning.)

SCORE: ____________

www.ingramcontent.com/pod-product-compliance
Lightning Source LLC
Chambersburg PA
CBHW032003190326
41520CB00007B/342
* 9 7 8 0 9 9 8 5 8 4 7 9 9 *